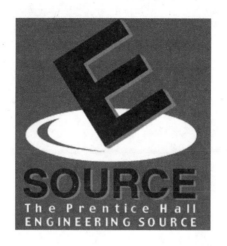

SOURCE
The Prentice Hall
ENGINEERING SOURCE

Graphics Concepts

Richard M. Lueptow
Northwestern University

Prentice Hall
Upper Saddle River, NJ 07458

Library of Congress Cataloging-in-Publication Data
Lueptow, Richard M.
    Graphics concepts / Richard M. Lueptow.
       p.   cm. — (ESource—the Prentice Hall engineering source)
    Includes bibliographical reference and index.
    ISBN 0–13–030687–8
    1. Engineering graphics. I. Title. II. Series.

    T385.D6327 2000
    620'.0022—dc21

                                                    00.029824

Vice-president of editorial development, ECS: **MARCIA HORTON**
Acquisitions editor: **ERIC SVENDSEN**
Associate editor: **JOE RUSSO**
Vice-president of production and manufacturing: **DAVID W. RICCARDI**
Executive managing editor: **VINCE O' BRIEN**
Managing editor: **DAVID A. GEORGE**
Editorial/production supervisor: **LAKSHMI BALASUBRAMANIAN**
Cover director: **JAYNE CONTE**
Manufacturing buyer: **PAT BROWN**
Editorial assistant: **KRISTEN BLANCO**
Market manager: **DANNY HOYT**

© 2000 by Prentice-Hall, Inc.
Upper Saddle River, New Jersey 07458

The author and publisher of this book have used their best efforts in
preparing this book. These efforts include the development, research,
and testing of the theories to determine their effectiveness.
Printed in the United States of America.

10 9 8 7 6 5 4 3 2 1

ISBN 0-13-030687-8

Prentice-Hall International (UK) Limited, *London*
Prentice-Hall of Australia Pty. Limited, *Sydney*
Prentice-Hall Canada, Inc., Toronto
Prentice-Hall Hispanoamericana, S.A., *Mexico*
Prentice-Hall of India Private Limited, *New Delhi*
Prentice-Hall of Japan, Inc., *Tokyo*
Pearson Education (Singapore) Pte. Ltd., *Singapore*
Editoria Prentice-Hall do Brasil, Ltda., *Rio de Janeiro*

About ESource

ESource—The Prentice Hall Engineering Source
—www.prenhall.com/esource

ESource—The Prentice Hall Engineering Source gives professors the power to harness the full potential of their text and their first-year engineering course. More than just a collection of books, ESource is a unique publishing system revolving around the ESource website—www.prenhall.com/esource. ESource enables you to put your stamp on your book just as you do your course. It lets you:

Control   You choose exactly what chapter or sections are in your book and in what order they appear. Of course, you can choose the entire book if you'd like and stay with the authors' original order.

Optimize   Get the most from your book and your course. ESource lets you produce the optimal text for your students needs.

Customize   You can add your own material anywhere in your text's presentation, and your final product will arrive at your bookstore as a professionally formatted text.

ESource ACCESS

Starting in the fall of 2000, professors who choose to bundle two or more texts from the ESource series for their class, or use an ESource custom book will be providing their students with complete access to the library of ESource content. All bundles and custom books will come with a student password that gives web ESource ACCESS to all information on the site. This passcode is free and is valid for one year after initial log-on. We've designed ESource ACCESS to provides students a flexible, searchable, on-line resource.

ESource Content

All the content in ESource was written by educators specifically for freshman/first-year students. Authors tried to strike a balanced level of presentation, an approach that was neither formulaic nor trivial, and one that did not focus too heavily on advanced topics that most introductory students do not encounter until later classes. Because many professors do not have extensive time to cover these topics in the classroom, authors prepared each text with the idea that many students would use it for self-instruction and independent study. Students should be able to use this content to learn the software tool or subject on their own.

　　　　While authors had the freedom to write texts in a style appropriate to their particular subject, all followed certain guidelines created to promote a consistency that makes students comfortable. Namely, every chapter opens with a clear set of **Objectives**, includes **Practice Boxes** throughout the chapter, and ends with a number of **Problems**, and a list of **Key Terms**. **Applications Boxes** are spread throughout the book with the intent of giving students a real-world perspective of engineering. **Success Boxes** provide the student with advice about college study skills, and help students avoid the common pitfalls of first-year students. In addition, this series contains an entire book titled *Engineering Success* by Peter Schiavone of the University of Alberta intended to expose students quickly to what it takes to be an engineering student.

Creating Your Book

Using ESource is simple. You preview the content either on-line or through examination copies of the books you can request on-line, from your PH sales rep, or by calling 1-800-526-0485. Create an on-line outline of the content you want, in the order you want, using ESource's simple interface. Either type or cut and paste your own material and insert it into the text flow. You can preview the overall organization of the text you've created at anytime (please note, since this preview is immediate, it comes unformatted.), then press another button and receive an order number for your own custom book. If you are not ready to order, do nothing—ESource will save your work. You can come back at any time and change, re-arrange, or add more material to your creation. You are in control. Once you're finished and you have an ISBN, give it to your bookstore and your book will arrive on their shelves six weeks after they order. Your custom desk copies with their instructor supplements will arrive at your address at the same time.

To learn more about this new system for creating the perfect textbook, go to www.prenhall.com/esource. You can either go through the on-line walkthrough of how to create a book, or experiment yourself.

Supplements

Adopters of ESource receive an instructor's CD that contains professor and student code from the books in the series, as well as other instruction aides provided by authors. The website also holds approximately **350 Powerpoint transparencies** created by Jack Leifer of Univ. of Kentucky—Paducah available to download. Professors can either follow these transparencies as pre-prepared lectures or use them as the basis for their own custom presentations.

Titles in the ESource Series

Introduction to UNIX
0-13-095135-8
David I. Schwartz

Introduction to AutoCAD 2000
0-13-016732-0
Mark Dix and Paul Riley

Introduction to Maple
0-13-095133-1
David I. Schwartz

Introduction to Word
0-13-254764-3
David C. Kuncicky

Introduction to Excel, 2/e
0-13-016881-5
David C. Kuncicky

Introduction to Mathcad
0-13-937493-0
Ronald W. Larsen

Introduction to AutoCAD, R. 14
0-13-011001-9
Mark Dix and Paul Riley

Introduction to the Internet, 3/e
0-13-031355-6
Scott D. James

Design Concepts for Engineers
0-13-081369-9
Mark N. Horenstein

Engineering Design—A Day in the Life of Four Engineers
0-13-085089-6
Mark N. Horenstein

Engineering Ethics
0-13-784224-4
Charles B. Fleddermann

Engineering Success
0-13-080859-8
Peter Schiavone

Mathematics Review
0-13-011501-0
Peter Schiavone

Introduction to C
0-13-011854-0
Delores Etter

Introduction to C++
0-13-011855-9
Delores Etter

Introduction to MATLAB
0-13-013149-0
Delores Etter with David C. Kuncicky

Titles in the ESource Series

Introduction to FORTRAN 90
0-13-013146-6
Larry Nyhoff and Sanford Lesetma

Introduction to Java
0-13-919416-9
Stephen J. Chapman

Introduction to Engineering Analysis
0-13-016733-9
Kirk D. Hagen

Introduction to PowerPoint
0-13-040214-1
Jack Leifer

Graphics Concepts
0-13-030687-8
Richard M. Lueptow

Graphics Concepts with Pro/ENGINEER
0-13-014154-2
Richard M. Lueptow, Jim Steger, and Michael T. Snyder

Graphics Concepts with SolidWorks
0-13-014155-0
Richard M. Lueptow and Michael Minbiole

Introduction to Visual Basic 6.0
0-13-026813-5
David I. Schneider

Introduction to Mathcad 2000
0-13-020007-7
Ronald W. Larsen

About the Authors

No project could ever come to pass without a group of authors who have the vision and the courage to turn a stack of blank paper into a book. The authors in this series worked diligently to produce their books, provide the building blocks of the series.

Delores M. Etter is a Professor of Electrical and Computer Engineering at the University of Colorado. Dr. Etter was a faculty member at the University of New Mexico and also a Visiting Professor at Stanford University. Dr. Etter was responsible for the Freshman Engineering Program at the University of New Mexico and is active in the Integrated Teaching Laboratory at the University of Colorado. She was elected a Fellow of the Institute of Electrical and Electronics Engineers for her contributions to education and for her technical leadership in digital signal processing. In addition to writing best-selling textbooks for engineering computing, Dr. Etter has also published research in the area of adaptive signal processing.

Sanford Leestma is a Professor of Mathematics and Computer Science at Calvin College, and received his Ph.D. from New Mexico State University. He has been the long-time co-author of successful textbooks on Fortran, Pascal, and data structures in Pascal. His current research interest are in the areas of algorithms and numerical computation.

Larry Nyhoff is a Professor of Mathematics and Computer Science at Calvin College. After doing bachelor's work at Calvin, and Master's work at Michigan, he received a Ph.D. from Michigan State and also did graduate work in computer science at Western Michigan. Dr. Nyhoff has taught at Calvin for the past 34 years—mathematics at first and computer science for the past several years. He has co-authored several computer science textbooks since 1981 including titles on Fortran and C++, as well as a brand new title on Data Structures in C++.

Acknowledgments: We express our sincere appreciation to all who helped in the preparation of this module, especially our acquisitions editor Alan Apt, managing editor Laura Steele, developmental editor Sandra Chavez, and production editor Judy Winthrop. We also thank Larry Genalo for several examples and exercises and Erin Fulp for the Internet address application in Chapter 10. We appreciate the insightful review provided by Bart Childs. We thank our families—Shar, Jeff, Dawn, Rebecca, Megan, Sara, Greg, Julie, Joshua, Derek, Tom, Joan; Marge, Michelle, Sandy, Lory, Michael—for being patient and understanding. We thank God for allowing us to write this text.

Mark Dix began working with AutoCAD in 1985 as a programmer for CAD Support Associates, Inc. He helped design a system for creating estimates and bills of material directly from AutoCAD drawing databases for use in the automated conveyor industry. This system became the basis for systems still widely in use today. In 1986 he began collaborating with Paul Riley to create AutoCAD training materials, combining Riley's background in industrial design and training with Dix's background in writing, curriculum development, and programming. Dix and Riley have created tutorial and teaching methods for every AutoCAD release since Version 2.5. Mr. Dix has a Master of Education from the University of Massachusetts. He is currently the Director of Dearborn Academy High School in Arlington, Massachusetts.

Paul Riley is an author, instructor, and designer specializing in graphics and design for multimedia. He is a founding partner of CAD Support Associates, a contract service and professional training organization for computer-aided design. His 15 years of business experience and 20 years of teaching experience are supported by degrees in education and computer science. Paul has taught AutoCAD at the University of Massachusetts at Lowell and is presently teaching AutoCAD at Mt. Ida College in Newton, Massachusetts. He has developed a program, Computer-aided Design for Professionals that is highly regarded by corporate clients and has been an ongoing success since 1982.

Scott D. James is a staff lecturer at Kettering University (formerly GMI Engineering & Management Institute) in Flint, Michigan. He is currently pursuing a Ph.D. in Systems Engineering with an emphasis on software engineering and computer-integrated manufacturing. Scott decided on writing textbooks after he found a void in the books that were available. "I really wanted a book that showed how to do things in good detail but in a clear and concise way. Many of the books on the market are full of fluff and force you to dig out the really important facts." Scott decided on teaching as a profession after several years in the computer industry. "I thought that it was really important to know what it was like outside of academia. I wanted to provide students with classes that were up to date and provide the information that is really used and needed."
Acknowledgments: Scott would like to acknowledge his family for the time to work on the text and his students and peers at Kettering who offered helpful critiques of the materials that eventually became the book.

Charles B. Fleddermann is a professor in the Department of Electrical and Computer Engineering at the University of New Mexico in Albuquerque, New Mexico. All of his degrees are in electrical engineering: his Bachelor's degree from the University of Notre Dame, and the Master's and Ph.D. from the University of Illinois at Urbana-Champaign. Prof. Fleddermann developed an engineering ethics course for his department in response to the ABET requirement to incorporate ethics topics into the undergraduate engineering curriculum. *Engineering Ethics* was written as a vehicle for presenting ethical theory, analysis, and problem solving to engineering undergraduates in a concise and readily accessible way.
Acknowledgments: I would like to thank Profs. Charles Harris and Michael Rabins of Texas A & M University whose NSF sponsored workshops on engineering ethics got me started thinking in this field. Special thanks to my wife Liz, who proofread the manuscript for this book, provided many useful suggestions, and who helped me learn how to teach "soft" topics to engineers.

David I. Schwartz is an Assistant Professor in the Computer Science Department at Cornell University and earned his B.S., M.S., and Ph.D. degrees in Civil Engineering from State University of New York at Buffalo. Throughout his graduate studies, Schwartz combined principles of computer science to applications of civil engineering. He became interested in helping students learn how to apply software tools for solving a variety of engineering problems. He teaches his students to learn incrementally and practice frequently to gain the maturity to tackle other subjects. In his spare time, Schwartz plays drums in a variety of bands.
Acknowledgments: I dedicate my books to my family, friends, and students who all helped in so many ways. Many thanks go to the schools of Civil Engineering and Engineering & Applied Science at State University of New York at Buffalo where I originally developed and tested my UNIX and Maple books. I greatly appreciate the opportunity to explore my goals and all the help from everyone at the Computer Science Department at Cornell. Eric Svendsen and everyone at Prentice Hall also deserve my gratitude for helping to make these books a reality. Many thanks, also, to those who submitted interviews and images.

Ron Larsen is an Associate Professor of Chemical Engineering at Montana State University, and received his Ph.D. from the Pennsylvania State University. He was initially attracted to engineering by the challenges the profession offers, but also appreciates that engineering is a serving profession. Some of the greatest challenges he has faced while teaching have involved non-traditional teaching methods, including evening courses for practicing engineers and teaching through an interpreter at the Mongolian National University. These experiences have provided tremendous opportunities to learn new ways to communicate technical material. He tries to incorporate the skills he has learned in non-traditional arenas to improve his lectures, written materials, and learning programs. Dr. Larsen views modern software as one of the new tools that will radically alter the way engineers work, and his book *Introduction to Mathcad* was written to help young engineers prepare to meet the challenges of an ever-changing workplace.
Acknowledgments: To my students at Montana State University who have endured the rough drafts and typos, and who still allow me to experiment with their classes—my sincere thanks.

Peter Schiavone is a professor and student advisor in the Department of Mechanical Engineering at the University of Alberta, Canada. He received his Ph.D. from the University of Strathclyde, U.K. in 1988. He has authored several books in the area of student academic success as well as numerous papers in international scientific research journals. Dr. Schiavone has worked in private industry in several different areas of engineering including aerospace and systems engineering. He founded the first Mathematics Resource Center at the University of Alberta, a unit designed specifically to teach new students the necessary *survival skills* in mathematics and the physical sciences required for success in first-year engineering. This led to the Students' Union Gold Key Award for outstanding contributions to the university. Dr. Schiavone lectures regularly to freshman engineering students and to new engineering professors on engineering success, in particular about maximizing students' academic performance. He wrote the book *Engineering Success* in order to share the *secrets of success in engineering study*: the most effective, tried and tested methods used by the most successful engineering students.

Acknowledgements: Thanks to Eric Svendsen for his encouragement and support; to Richard Felder for being such an inspiration; to my wife Linda for sharing my dreams and believing in me; and to Francesca and Antonio for putting up with Dad when working on the text.

Mark N. Horenstein is a Professor in the Department of Electrical and Computer Engineering at Boston University. He has degrees in Electrical Engineering from M.I.T. and U.C. Berkeley and has been involved in teaching engineering design for the greater part of his academic career. He devised and developed the senior design project class taken by all electrical and computer engineering students at Boston University. In this class, the students work for a virtual engineering company developing products and systems for real-world engineering and social-service clients. Many of the design projects developed in his class have been aimed at assistive technologies for individuals with disabilities.

Acknowledgments: I would like to thank Prof. James Bethune, the architect of the Peak Performance event at Boston University, for his permission to highlight the competition in my text. Several of the ideas relating to brainstorming and teamwork were derived from a workshop on engineering design offered by Prof. Charles Lovas of Southern Methodist University. The principles of estimation were derived in part from a freshman engineering problem posed by Prof. Thomas Kincaid of Boston University.

Kirk D. Hagen is a professor at Weber State University in Ogden, Utah. He has taught introductory-level engineering courses and upper-division thermal science courses at WSU since 1993. He received his B.S. degree in physics from Weber State College and his M.S. degree

in mechanical engineering from Utah State University, after which he worked as a thermal designer/analyst in the aerospace and electronics industries. After several years of engineering practice, he resumed his formal education, earning his Ph.D. in mechanical engineering at the University of Utah. Hagen is the author of an undergraduate heat transfer text. Having drawn upon his industrial and teaching experience, he strongly believes that engineering students must develop effective analytical problem solving abilities. His book, *Introduction to Engineering Analysis*, was written to help beginning engineering students learn a systematic approach to engineering analysis.

Richard M. Lueptow is the Charles Deering McCormick Professor of Teaching Excellence and Associate Professor of Mechanical Engineering at Northwestern University. He is a native of Wisconsin and received his doctorate from the Massachusetts Institute of Technology in 1986. He teaches design, fluid mechanics, and spectral analysis techniques. "In my design class I saw a need for a self-paced tutorial for my students to learn CAD software quickly and easily. I worked with several students a few years ago to develop just this type of tutorial, which has since evolved into a book. My goal is to introduce students to engineering graphics and CAD, while showing them how much fun it can be." Rich has an active research program on rotating filtration, Taylor Couette flow, granular flow, fire suppression, and acoustics. He has five patents and over 40 refereed journal and proceedings papers along with many other articles, abstracts, and presentations.

Acknowledgments: Thanks to my talented and hard-working co-authors as well as the many colleagues and students who took the tutorial for a "test drive." Special thanks to Mike Minbiole for his major contributions to Graphics Concepts with SolidWorks. Thanks also to Northwestern University for the time to work on a book. Most of all, thanks to my loving wife, Maiya, and my children, Hannah and Kyle, for supporting me in this endeavor. (Photo courtesy of Evanston Photographic Studios, Inc.)

Jack Leifer is an Assistant Professor in the Department of Mechanical Engineering at the University of Kentucky Extended Campus Program in Paducah, and was previously with the Department of Mathematical Sciences and Engineering at the University of South Carolina— Aiken. He received his Ph.D. in Mechanical Engineering from the University of Texas at Austin in December 1995.

His current research interests include the modeling of sensors for manufacturing, and the use of Artificial Neural Networks to predict corrosion.

Acknowledgements: I'd like to thank my colleagues at USC—Aiken, especially Professors Mike May and Laurene Fausett, for their encouragement and feedback; Eric Svendsen and Joe Russo of Prentice Hall, for their useful suggestions and flexibility with deadlines; and my parents, Felice and Morton Leifer, for being there and providing support (as always) as I completed this book.

David C. Kuncicky is a native Floridian. He earned his Baccalaureate in psychology, Master's in computer science, and Ph.D. in computer science from Florida State University. He has served as a faculty member in the Department of Electrical Engineering at the FAMU–FSU College of Engineering and the Department of Computer Science at Florida State University. He has taught computer science and computer engineering courses for over 15 years. He has published research in the areas of intelligent hybrid systems and neural networks. He is currently the Director of Engineering at Bioreason, Inc. in Sante Fe, New Mexico.

Acknowledgments: Thanks to Steffie and Helen for putting up with my late nights and long weekends at the computer. Thanks also to the helpful and insightful technical reviews by Jerry Ralya, Kathy Kitto, Avi Singhal, Thomas Hill, Ron Eaglin, Larry Richards, and Susan Freeman. I appreciate the patience of Eric Svendsen and Joe Russo of Prentice Hall for gently guiding me through this project. Finally, thanks to Susan Bassett for having faith in my abilities, and for providing continued tutelage and support.

Jim Steger is currently Chief Technical Officer and cofounder of an Internet applications company. He graduated with a Bachelor of Science degree in Mechanical Engineering from Northwestern University. His prior work included mechanical engineering assignments at Motorola and Acco Brands. At Motorola, Jim worked on part design for two-way radios and was one of the lead mechanical engineers on a cellular phone product line. At Acco Brands, Jim was the sole engineer on numerous office product designs. His Worx stapler has won design awards in the United States and in Europe. Jim has been a Pro/Engineer user for over six years.

Acknowledgments: Many thanks to my co-authors, especially Rich Lueptow for his leadership on this project. I would also like to thank my family for their continuous support.

David I. Schneider holds an A.B. degree from Oberlin College and a Ph.D. degree in Mathematics from MIT. He has taught for 34 years, primarily at the University of Maryland. Dr. Schneider has authored 28 books, with one-half of them computer programming books. He has developed three customized software packages that are supplied as supplements to over 55 mathematics textbooks. His involvement with computers dates back to 1962, when he programmed a special purpose computer at MIT's Lincoln Laboratory to correct errors in a communications system.

Michael T. Snyder is President of Internet startup Appointments123.com. He is a native of Chicago, and he received his Bachelor of Science degree in Mechanical Engineering from the University of Notre Dame. Mike also graduated with honors from Northwestern University's Kellogg Graduate School of Management in 1999 with his Masters of Management degree. Before Appointments123.com, Mike was a mechanical engineer in new product development for Motorola Cellular and Acco Office Products. He has received four patents for his mechanical design work. "Pro/Engineer was an invaluable design tool for me, and I am glad to help students learn the basics of Pro/Engineer."

Acknowledgments: Thanks to Rich Lueptow and Jim Steger for inviting me to be a part of this great project. Of course, thanks to my wife Gretchen for her support in my various projects.

Stephen J. Chapman received a BS in Electrical Engineering from Louisiana State University (1975), an MSE in Electrical Engineering from the University of Central Florida (1979), and pursued further graduate studies at Rice University. Mr. Chapman is currently Manager of Technical Systems for Brithish Aerospace Australia, in Melbourne, Australia. In this position, he provides technical direction and design authority for the work of younger engineers within the company. He is also continuing to teach at local universities on a part-time basis.

Mr. Chapman is a Senior Member of the Institute of Electrical and Electronics Engineers (and several of its component societies). He is also a member of the Association for Computing Machinery and the Institution of Engineers (Australia).

Reviewers

ESource benefited from a wealth of reviewers who on the series from its initial idea stage to its completion. Reviewers read manuscripts and contributed insightful comments that helped the authors write great books. We would like to thank everyone who helped us with this project.

Concept Document

Naeem Abdurrahman *University of Texas, Austin*
Grant Baker *University of Alaska, Anchorage*
Betty Barr *University of Houston*
William Beckwith *Clemson University*
Ramzi Bualuan *University of Notre Dame*
Dale Calkins *University of Washington*
Arthur Clausing *University of Illinois at Urbana –Champaign*
John Glover *University of Houston*
A.S. Hodel *Auburn University*
Denise Jackson *University of Tennessee, Knoxville*
Kathleen Kitto *Western Washington University*
Terry Kohutek *Texas A&M University*
Larry Richards *University of Virginia*
Avi Singhal *Arizona State University*
Joseph Wujek *University of California, Berkeley*
Mandochehr Zoghi *University of Dayton*

Books

Stephen Allan *Utah State University*
Naeem Abdurrahman *University of Texas, Austin*
Anil Bajaj *Purdue University*
Grant Baker *University of Alaska—Anchorage*
Betty Burr *University of Houston*
William Beckwith *Clemson University*
Haym Benaroya *Rutgers University*
Tom Bledsaw *ITT Technical Institute*
Tom Bryson *University of Missouri, Rolla*
Ramzi Bualuan *University of Notre Dame*
Dan Budny *Purdue University*
Dale Calkins *University of Washington*
Arthur Clausing *University of Illinois*
James Devine *University of South Florida*

Patrick Fitzhorn *Colorado State University*
Dale Elifrits *University of Missouri, Rolla*
Frank Gerlitz *Washtenaw College*
John Glover *University of Houston*
John Graham *University of North Carolina—Charlotte*
Malcom Heimer *Florida International University*
A.S. Hodel *Auburn University*
Vern Johnson *University of Arizona*
Kathleen Kitto *Western Washington University*
Robert Montgomery *Purdue University*
Mark Nagurka *Marquette University*
Romarathnam Narasimhan *University of Miami*
Larry Richards *University of Virginia*
Marc H. Richman *Brown University*
Avi Singhal *Arizona State University*
Tim Sykes *Houston Community College*
Thomas Hill *SUNY at Buffalo*
Michael S. Wells *Tennessee Tech University*
Joseph Wujek *University of California, Berkeley*
Edward Young *University of South Carolina*
Mandochehr Zoghi *University of Dayton*
John Biddle *California State Polytechnic University*
Fred Boadu *Duke University*
Harish Cherukuri *University of North Carolina –Charlotte*
Barry Crittendon *Virginia Polytechnic and State University*
Ron Eaglin *University of Central Florida*
Susan Freeman *Northeastern University*
Frank Gerlitz *Washtenaw Community College*
Otto Gygax *Oregon State University*
Donald Herling *Oregon State University*
James N. Jensen *SUNY at Buffalo*
Autar Kaw *University of South Florida*
Kenneth Klika *University of Akron*
Terry L. Kohutek *Texas A&M University*
Melvin J. Maron *University of Louisville*
Soronadi Nnaji *Florida A&M University*
Michael Peshkin *Northwestern University*
Randy Shih *Oregon Institute of Technology*
Neil R. Thompson *University of Waterloo*
Garry Young *Oklahoma State University*

Contents

ABOUT ESOURCE   v

ABOUT THE AUTHORS   ix

1 ENGINEERING GRAPHICS   1

1.1   The Importance of Engineering Graphics   1
1.2   Engineering Graphics   3
1.3   CAD   6
1.4   Design and CAD   7

2 PROJECTIONS USED IN ENGINEERING GRAPHICS   11

2.1   Projections   11
2.2   3-D Projections   12
2.3   Multiview Projections   14
2.4   Working Drawings   17

3 FREEHAND SKETCHING   23

3.1   Why Freehand Sketches?   23
3.2   Freehand Sketching Fundamentals   24
3.3   Basic Freehand Sketching   25
         3.3.1   Oblique Sketching   26
         3.3.2   Isometric Sketching   27
3.4   Advanced Freehand Sketching   28
         3.4.1   Freehand Oblique Sketching   28
         3.4.2   Isometric Sketching   30
         3.4.3   Orthographic Sketching   32

4 COMPUTER AIDED DESIGN AND DRAFTING   37

4.1   CAD Models   37
4.2   CAD and Solids Modeling   40
4.3   The Nature of Solids Modeling   42

5 STANDARD PRACTICE FOR ENGINEERING DRAWINGS   49

5.1   Introduction to Drawing Standards   49
5.2   Sheet Layouts   50
5.3   Lines   51
5.4   Dimension Placement and Conventions   52
5.5   Section and Detail Views   56
5.6   Fasteners and Screw Threads   59
5.7   Assembly Drawings   61

6 TOLERANCES   67

6.1   Why Tolerances?   67
6.2   Displaying Tolerances on Drawings   69
6.3   How to Determine Tolerances   70
6.4   Surface Finish   73
6.5   Geometric Dimensioning and Tolerancing   74

1

Engineering Graphics

SECTIONS

- 1.1 The Importance of Engineering Graphics
- 1.2 Engineering Graphics
- 1.3 CAD
- 1.4 Design and CAD

OVERVIEW

Engineering designs start as images in the mind's eye of an engineer. Engineering graphics has evolved to communicate and record these ideas on paper both two- and three-dimensionally. In the past few decades, the computer has made it possible to automate the creation of engineering graphics. Today engineering design and engineering graphics are inextricably connected. Engineering design is communicated visually using engineering graphics.

OBJECTIVES

After reading this chapter, you should be able to:

- Describe visual thinking
- Differentiate perspective, isometric, and orthographic projections
- Understand the basis of CAD
- Understand the relationship between design and CAD

1.1 THE IMPORTANCE OF ENGINEERING GRAPHICS

"Visualizing" a picture or image in your mind is a familiar experience. The image can be visualized at many different levels of abstraction. Think about light and you might see the image of a light bulb in your "mind's eye." Alternatively, you might think about light versus dark. Or you might visualize a flashlight or table lamp. Such visual thinking is necessary in engineering and science. Albert Einstein said that he rarely thought in words. Instead, he laboriously translated his visual images into verbal and mathematical terms.

Visual thinking is a foundation of engineering. Walter P. Chrysler, founder of the automobile company, recounted his experience as an apprentice machinist where he built a model locomotive that existed "within my mind so real, so complete, that it seemed to have three dimensions there."

Yet, the complexity of today's technology rarely permits a single person to build a device from his own visual image. The images must be conveyed to other engineers and designers. In addition, those images must be constructed in such a way that they are in a readily recognizable, consistent, and readable format. This assurress that the visual ideas are clearly and unambiguously conveyed to others. *Engineering graphics* is a highly stylized way of presenting images of parts or assemblies.

A major portion of engineering information is recorded and transmitted using engineering graphics. In fact, 92 percent of the design process is graphically based. Written and verbal communications along with mathematics account for the remaining eight percent. To demonstrate the effectiveness of engineering graphics compared to a written description try to visualize an ice scraper based on this word description:

> An ice scraper is generally in the shape of a 140 × 80 × 10 mm rectangular prism. One end is beveled from zero thickness to the maximum thickness in a length of 40 mm to form a sharp edge. The opposite end is semicircular. A 20 mm diameter hole is positioned so the center of the hole is 40 mm from the semicircular end and 40 mm from either side of the scraper.

It is evident immediately that the shape of the ice scraper is much more easily visualized from the graphical representation shown in Figure 1.1 than from the word description. Humans grasp information much more quickly when that information is presented in a graphical or visual form rather than as a word description.

Figure 1.1.

Engineering drawings, whether done using a pencil and paper or a computer, start with a blank page or screen. The engineer's mind's eye image must be transferred to the paper or computer screen. The creative nature of this activity is similar to that of an artist. Perhaps the greatest example of this is Leonardo da Vinci, who had exceptional engineering creativity devising items such as parachutes and ball bearings, shown in Figure 1.2, hundreds of years before they were re-invented. He also had exceptional artistic talent, creating some of the most famous pictures ever painted such as *Mona Lisa* and *The Last Supper*.

Figure 1.2. The creative genius of da Vinci is evident in these sketches of a parachute and a ball bearing, both devised hundreds of years before they were re-invented. (Parachute used with permission of the Biblioteca–Pinacoteca Ambrosiana, Milan, Italy. Property of the Ambrosian Library. All rights reserved. Reproduction is forbidden. [*The Inventions of Leonardo da Vinci*, Charles Gibbs-Smith, Phaidon Press, Oxford, 1978, p. 24.] Ball bearing used with permission of EMB-Service for Publishers, Lucerne, Switzerland. [*The Unknown Leonardo*, edited by Ladislao Reti, McGraw-Hill, 1974, p. 286])

1.2 ENGINEERING GRAPHICS

The first authentic record of engineering graphics dates back to 2130 BC, based on a statue now in the Museum of the Louvre, Paris. The statue depicts an engineer and governor of a small city-state in an area later known as Babylon. At the base of the statue are measuring scales and scribing instruments along with a plan of a fortress engraved on a stone tablet.

Except for the use of pen and paper rather than stone tablets, it was not until printed books appeared around 1450 that techniques of graphics advanced. Around the same time, *pictorial perspective* drawing was invented by artist Paolo Uccello. This type of drawing presents an object much like it would look to the human eye or in a photograph, as shown in Figure 1.3. The essential characteristic of a perspective drawing is that parallel lines converge at a point in the distance like parallel railroad tracks seem to converge in the distance. Copper-plate engravings permitted the production of finely detailed technical drawings using pictorial perspective in large numbers. The pictorial perspective drawings were crucial to the advancement of technology through the Renaissance and until the beginning of the Industrial Revolution. But these drawings could not convey adequately details of the construction of an object. One solution to this problem was the use of the *exploded view* developed in the 15th century and perfected by Leonardo da Vinci. The exploded view of an assembly of individual parts shows the parts spread out along a common axis, as shown for the hoist in Figure 1.4. The exploded view reveals details of the individual parts along with showing the order in which they are assembled.

Figure 1.3. An example of pictorial perspective by Agostino Ramelli in 1588. (Used with permission of the Syndics of Cambridge University Library. [*The Various and Ingenious Machines of Agostino Ramelli (1588),* translated by Martha Teach Gnudi, Johns Hopkins University Press, 1976. p. 83.])

The Industrial Revolution brought with it the need to tie more closely the concept of a design with the final manufactured product using technical drawing. The perspective drawing of a simple object in Figure 1.5a shows pictorially what the object looks like. However, it is difficult to represent accurately dimensions and other details in a perspective drawing. *Orthographic projections*, developed in 1528 by German artist Albrecht Dürer, accomplish this quite well. An orthographic projection typically shows three views of an object. Each view shows a different side of the object (say the front, top, and side). An example of an orthographic projection is shown in Figure 1.5b. Orthographic projections are typically easy to draw, and the lengths and angles in orthographic projections have little distortion. As a result, orthographic drawings can convey more information than a perspective drawing. But their interpretation takes more effort than a pictorial perspective, as is evident from Figure 1.5. French philosopher and mathematician René Descartes laid the foundation for the mathematical principles of projections by connecting geometry to algebra in the 17th century. Much later Gaspard Monge, a French mathematician, "invented" the mathematical principles

Figure 1.4. Assembled and exploded view of a hoist by Leonardo da Vinci, circa 1500. (Used with permission of the Biblioteca-Pinacoteca Ambrosiana, Milan, Italy. Property of the Ambrosian Library. All rights reserved. Reproduction is forbidden. [*The Inventions of Leonardo da Vinci*, Charles Gibbs-Smith, Phaidon Press, Oxford, 1978, p. 64.])

of projection known as *descriptive geometry*. These principles form the basis of engineering graphics today. But because these principles were thought to be of such strategic importance, they remained military secrets until 1795. By the 19th century, orthographic projections were used almost universally in mechanical drawing, and they are still the basis for engineering drawings today.

The *isometric view* is used more often today than the pictorial perspective view. Historically, it was used for centuries by engravers. Then in the early 19th century, William Farish, an English mathematician, formalized the isometric view and introduced it to engineers. The isometric view simplifies the pictorial perspective. In an isometric view, parallel lines remain parallel rather than converging to a point in the distance, as shown in Fig. 1.5c. Keeping parallel lines parallel distorts the appearance of the object slightly. But the distortion in an isometric view is negligible for objects of limited depth. For situations in which the depth of the object is large, such as an architectural view down a long hallway, the pictorial perspective is preferable. The advantage of the isometric view, though, is that it is much easier to draw than a pictorial perspective view.

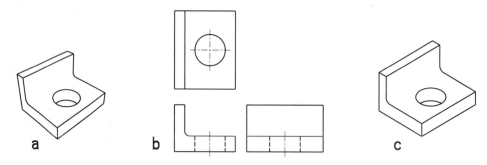

a   b   c

Figure 1.5.

1.3   CAD

The introduction of the computer revolutionized engineering graphics. Pioneers in computer-aided engineering graphics envisioned the computer as a tool to replace paper and pencil drafting with a system that is more automated, efficient, and accurate. The first demonstration of a computer-based drafting tool was a system called SKETCHPAD developed at the Massachusetts Institute of Technology in 1963 by Ivan Sutherland. The system used a monochrome monitor with a light pen for input from the user. The following year IBM commercialized computer-aided drafting.

During the 1970s, computer-aided drafting blossomed as the technology changed from scientific endeavor to an economically indispensable industrial tool for design. Commands for geometry generators to create commonly occurring shapes were added. Functions were added to control the viewing of the drawing geometry. Modifiers such as rotate, delete, and mirror were implemented. Commands could be accessed by typing on the keyboard or by using a mouse. Perhaps most importantly, three-dimensional modeling techniques became a key part of engineering graphics software.

By the 1980s, computer-aided drafting became fully developed in the marketplace as a standard tool in industry. In addition, the current technology of solids modeling came about. Solids models represent objects in the virtual environment of the computer just as they exist in reality, having a volume as well as surfaces and edges. The introduction of Pro/ENGINEER® in 1988 and SolidWorks® in the 1990s revolutionized computer-aided design and drafting. Today solids modeling remains the state-of-the-art technology.

What we have been referring to as computer-aided drafting is usually termed *CAD*, an acronym for Computer Aided Design, Computer Aided Drafting, or Computer Aided Design and Drafting. Originally the term Computer Aided Design included any technique that uses computers in the design process including drafting, stress analysis, and motion analysis. But over the last 35 years CAD has come to refer more specifically to Computer Aided Design and Drafting. Computer Aided Engineering (CAE) is used to refer to the broader range of computer-related design tools.

1.4 DESIGN AND CAD

Inextricably connected with engineering graphics is the *design process* in which an engineering or design team faces a particular engineering problem and devises a solution to that problem. Often the design team includes persons responsible for engineering, product design, production system design, manufacturing, marketing, and sales. The idea is to simultaneously develop the product and the manufacturing process for the product. This is known as *concurrent engineering* or *integrated product and process design*. Key to the success of concurrent engineering is communication of information. Engineering graphics is one of the primary methods used by concurrent engineering teams to record and transfer information during the design process.

The process of bringing a product to market is shown in Figure 1.6. The process begins with the identification of a market, or user need. After this, the design process is the key to bridging the gap between a user need and the manufacture and sales of a product. The design process can be broken down into three parts as indicated in Figure 1.6.

Figure 1.6.

The first part is the *specification* of the problem. The *design specification* is a list of requirements that the final product must meet, including size, performance, weight, and so on. The second part of the design process is called *ideation* or conceptual design. In this phase, the design team devises as many ideas for solutions to the design problem as possible and then narrows them down to the best one based on the specification. In many cases, a designer or engineer will quickly sketch ideas to explore or communicate the design concept to the rest of the design team. These freehand sketches form the basis for the details of the design that are laid down in the third phase of the design process. It is in this last phase of the *design process*, known as *detail design*, that CAD is crucial. The conceptual ideas for the product that were seen in a designer's mind's eye or are on paper as rough sketches must then be translated into the visual language of engineering graphics. In this way, the ideas can be understood clearly and accurately by the design team and other designers, engineers, fabricators, suppliers, and machinists. It is in this phase of the design that the nitty gritty details have to be worked out. Should the device be 30 mm long or should it be 35 mm long? How will one part fit with another? What size hole should be used? What material should be used? What manufacturing process will be used to make the part? The number of individual decisions that need to be made can be very large, even for a relatively simple part. Once the engineering drawings have been created, the product can be manufactured and eventually sold.

The use of CAD has had a great impact on the design process. For example, a part may be modified several times to meet the design specification or to mate with another part. Before the advent of CAD, these modifications were very tedious, time-consuming, and prone to error. However, CAD has made it possible to make these changes relatively easily and quickly. The connectivity of computers using local area networks then makes the revised electronic drawings available to a team of engineers at an instant. This is crucial as engineering systems become more complex and operational requirements become more stringent. For example, a modern jet aircraft has several million individual parts that must all fit together and perform safely for several decades.

Although CAD had a great impact on making the design process speedier and more accurate, the capabilities of the first few generations of CAD were still limited. Early CAD systems only provided a means of automating the drafting process to create orthographic engineering drawings. The designer or engineer would simply generate a line on the computer screen rather than drawing the line on paper. Current computer graphics software such as "paint" or "draw" programs for personal computers work this way. As CAD became more sophisticated, it helped automate the drafting process based on the "intelligence" of the software. CAD software lacking such intelligence required an engineer to draw a pair of parallel lines an exact distance apart by specifying coordinates of the endpoints of the lines. More advanced generations of CAD software permitted an engineer to draw approximately parallel lines using a mouse. Then the engineer would specify a particular distance between the lines and that the lines should be parallel. The CAD software then automatically placed these lines the specified distance apart and made them parallel. However, the major problem with the early generations of CAD software was that the designer or engineer was simply creating two-dimensional orthographic views of a three-dimensional part using a computer instead of a pencil and paper. From these two-dimensional views, the engineer still needed to reconstruct the mind's eye view of the three-dimensional image in the same way as if the drawings were created by hand.

The current generation of CAD software has had a very profound effect on the design process, because it is now possible to create a virtual prototype of a part or assembly on a computer. For example, the solids model of a pizza cutter is shown in Figure 1.7. Rather than translating a three-dimensional image from the mind's eye to a two-dimensional orthographic projection of the object, current CAD software starts with generating a three-dimensional virtual model of the object directly on the computer. This virtual model can be rotated so that it can be viewed from different angles. Several parts can be virtually assembled on the computer to make sure that they fit together. The assembled parts can be viewed as an assembly or in an exploded view. All of this is done in a virtual environment on the computer before the two-dimensional orthographic engineering drawings are even produced. It is still usually necessary to produce the orthographic engineering drawings. But these drawings only serve as a standard means of engineering graphics communication, rather than as a tedious, time-consuming task necessary to proceed with the design process.

Figure 1.7.

What Happened to Pencil and Paper Drawings?

CAD drawings have already replaced pencil and paper drawings. Through the 1970s and even into the 1980s many engineering and design facilities consisted of rows and rows of drafting tables with a designer or engineer hunched over a drawing on each table. Engineering colleges and universities required a full-year course in "engineering drafting" or "graphics communication" for all engineering students. Many of these students purchased a set of drawing instruments along with their first semester textbooks. They spent endless hours practicing lettering and drawing perfect circles.

Now nearly all of the drafting tables and drafting courses have been replaced by CAD. One can still find a drafting table here and there, but it is not for creating engineering drawings. Most often it is used for displaying a large CAD drawing to designers and engineers. They can make notes on the drawing or freehand sketch modifications on the drawing. The pencil is still an ideal means for generating ideas and quickly conveying those ideas to others. But eventually, all of the pencil markings are used as the basis for modifying the CAD drawing.

In some companies with traditional products that have not changed for decades (like a spoon or a chair), pencil and paper drawings are gradually being converted into electronic form. In some cases, the original drawing is simply scanned to create an electronic version. The scanned drawing cannot be modified, but the electronic version takes much less storage space than a hard copy. In other cases, the original pencil and paper drawings are being systematically converted to CAD drawings, so that they can be modified if necessary. In any case, though, the pencil and paper drawing is now just a part of the history of engineering.

KEY TERMS

CAD	Detail Design	Isometric view
Descriptive geometry	Engineering graphics	Orthographic projection
Design process	Exploded view	Pictorial perspective view
Design specification	Ideation	

Problems

1. Research and report on an important historical figure in engineering graphics such as Leonardo da Vinci, Albrecht Dürer, Gaspard Monge, René Descartes, William Farish, M. C. Escher, Frank Lloyd Wright, Ivan Sutherland, or Paolo Uccello.

2. Using the World Wide Web, search for computer-aided design web sites. Report on the variety of software available for CAD and its capabilities.

3. Trace the early development of CAD and report on your findings. A particularly useful starting point is an article by S. H. Clauser in the November 1981 issue of *Mechanical Engineering*.

4. Computer hardware and user interface development has had a profound effect on the evolution of CAD software. Research and report on the impact of the light pen, the graphics tablet, the direct view storage tube, raster graphics technology, the work-station computer, the personal computer, or the mouse on the capability and evolution of CAD. (Many textbooks on engineering graphics discuss these items.)

5. Outline the design process and indicate how the three steps in the design process would relate to:

   a. the design of a suspension system for a mountain bike.
   b. the design of an infant car seat.
   c. the design of headphones for a portable cassette player.
   d. the design of a squirrel-proof bird feeder.
   e. the design of a car-mounted bicycle carrier.
   f. the design of children's playground equipment.

6. Explain how "paint" or "draw" programs such as McDraw, Adobe Illustrator, MS Paint, or drawing tools in MS Word differ from CAD engineering graphics software.

2

Projections Used
in Engineering Graphics

OVERVIEW

Engineering graphics is a highly stylized scheme to represent three-dimensional objects on a two-dimensional paper or computer screen. This can be accomplished by representing all three dimensions of an object in a single image or by presenting a collection of views of different sides of the object. Working drawings are the practical result of using engineering graphics to represent objects.

2.1   PROJECTIONS

The goal in engineering graphics, whether it is freehand sketching or CAD, is to represent a physical object or the mind's eye image of an object so that the image can be conveyed to other persons. Objects can be shown as *3-D projections* or *multiview projections*. Figure 2.1 shows the handle of a pizza cutter shown in both ways. The 3-D projection clearly suggests the three-dimensional character of the handle, even though it is displayed on a two-dimensional medium (the page). 3-D projections are useful in that they provide an image that is similar to the image in the designer's mind's eye. But 3-D projections are often weak in providing adequate details of the object, and there is often some distortion of the object. For instance, a circular hole at the right end of the handle becomes an ellipse in an isometric 3-D projection.

Multiview projections are used to overcome the weaknesses of 3-D projections. Multiview projections are a collection of flat 2-D drawings of the different sides of an object.

SECTIONS

- 2.1   Projections
- 2.2   3-D Projections
- 2.3   Multiview Projections
- 2.4   Working Drawings

OBJECTIVES

After reading this chapter, you should:

- Understand the spatial relation between 3-D projections and multiview projections
- Be able to differentiate isometric, trimetric, perspective, and oblique 3-D projections
- Understand how multiview projections are related to the views of the sides of an object
- Know the proper placement of orthographic views
- Know the difference between third-angle and first-angle orthographic projections
- Be able to differentiate the various kinds of working drawings

For instance, the side and bottom end of the pizza cutter handle are shown in the multiview projection in Figure 2.1. Because there are two views, it is quite easy to depict details of the object. In addition, taken together, multiview projections provide a more accurate representation of the object than the 3-D projection—a circular hole appears as a circle in a multiview projection. On the other hand, multiview projections require substantial interpretation, and the overall shape of an object is often not obvious upon first glance. Consequently, the combination of the overall image provided by 3-D projections and details provided by multiview projections yield a representation of an object that is best. The shape of the object is immediately evident from the 3-D projection, and the detail needed for an accurate description of the object is available from the multiview projection.

Figure 2.1.

2.2   3-D PROJECTIONS

Three different types of 3-D projections are available in most CAD software: isometric, trimetric, and perspective. These three views of a cube are shown in Figure 2.2. In all three cases, these 3-D projections represent all three dimensions of the cube in a single planar image. Although it is clear in all three cases that the object is a cube, each type of 3-D projection has its advantages and disadvantages.

The *isometric* projection has a standard orientation that makes it the typical projection used in CAD. In an isometric projection, the width and depth dimensions are sketched at 30° above horizontal as shown in Figure 2.2a. This results in the three angles at the upper front corner of the cube being equal to 120°. The three sides of the cube are also equal, leading to the term iso (equal) -metric (measure). Isometric drawings work quite well for objects of limited depth. However, an isometric drawing distorts the object when the depth is significant. In this case, a pictorial perspective drawing is better.

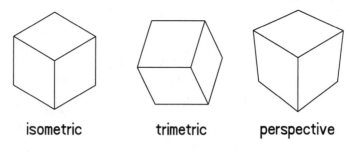

isometric   trimetric   perspective

Figure 2.2.

In general, the *trimetric* projection offers more flexibility in orienting the object in space. The width and depth dimensions are at arbitrary angles to the horizontal, and the three angles at the upper front corner of the cube are unequal. This makes the three sides of the cube each have a different length as measured in the plane of the drawing; hence the name tri-metric. In most CAD software, the trimetric projection fixes one side along a horizontal line and tips the cube forward as shown in Figure 2.2b. A *dimetric* projection sets two sides of the cube, usually those of the front face, equal.

A pictorial perspective, or simply *perspective*, projection is drawn so that parallel lines converge in the distance as shown in Figure 2.2c, unlike isometric or trimetric projections where parallel lines remain parallel. A perspective projection is quite useful in providing a realistic image of an object when the object spans a long distance, such as the view of a bridge or aircraft from one end. Generally, small manufactured objects are adequately represented by isometric or trimetric views.

Two types of pictorial sketches are used frequently in freehand sketching: isometric and oblique. The isometric projection was discussed with respect to 3-D CAD projections. The isometric projection is often used in freehand sketching because it is relatively easy to create a realistic sketch of an object. But the *oblique projection* is usually even easier to sketch. The oblique projection places the principal face of the object parallel to the plane of the paper with the axes in the plane of the paper perpendicular to one another. The axis into the paper is at an arbitrary angle with respect to the horizontal. Figure 2.3 compares an isometric projection and an oblique projection of a cube with a hole in it. The advantage of the oblique projection is that details in the front face of the object retain their true shape. For instance, the circle on the front face is circular in the oblique projection, while it is elliptical in the isometric projection. This feature often makes oblique freehand sketching somewhat easier than isometric sketching.

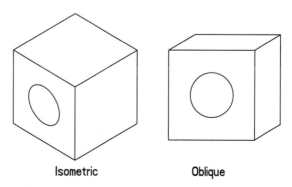

Isometric   Oblique

Figure 2.3.

2.3   MULTIVIEW PROJECTIONS

The standard means of multiview projection in engineering graphics is what we have referred to earlier as the *orthographic projection*. Although 3-D projections provide a readily identifiable visual image of an object, multiview projections are ideal for showing the details of an object. Dimensions can be shown easily and most features remain undistorted in multiview projections.

An orthographic projection is most easily thought of as a collection of views of different sides of an object—front, top, side, and so forth. For instance, two orthographic projections could be used for a coffee mug. The front view would show the sidewall of the mug along with the loop forming the handle. The top view would show what one would see looking down into the mug—a circular rim of the mug, the bottom of the inside of the mug, and the top of the handle that sticks out of the side of the mug. Dimensions of the mug could easily be added to the projections of each side of the mug to create an engineering drawing.

One useful way of looking at multiview projections is to imagine a glass box surrounding the object as shown in Figure 2.4. The image of each side of the object can be projected onto the wall of the glass box. Now an observer on the outside of the box can

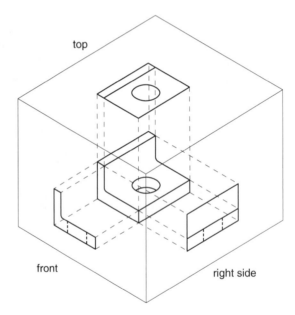

Figure 2.4.

see each side of the object as projected on each of the six walls of the box. Solid lines show the edges evident in the projection, and dashed lines show lines that are hidden by the object. Now imagine unfolding the glass box as if each of the edges of the glass box were a hinge, so that the front view is in the middle. Now the unfolded glass box represents all six sides of the object in a single plane as shown in Figure 2.5. In unfolding the glass box, the top view is positioned above the front view, the bottom view is below the front view, the right-side view is to the right of the front view, and so on. The dimensions of the object remain the same in all views. For example, the horizontal dimension (width of the object) in the front view is identical in that same dimension in the top view and bottom view. The views also remain aligned so that the bottom edge in the front

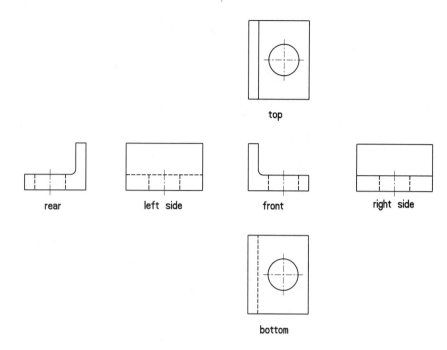

top

rear   **left side**   **front**   **right side**

bottom

Figure 2.5.

view is even with the bottom edge in the right side, left side, and rear views. Likewise, the top edges remain aligned. Finally, the same edges in adjacent views are closest together. For instance, the same edge of the object is at the left-side of the front view and the right side of the left side view. This edge in the front view is closest to the same edge in the left-side view.

In many cases, three views are needed to represent an object accurately, although in some cases (like a coffee mug) only two views are necessary, and in other cases more than three views are needed to show complex features of the object. It is helpful to select the side of the object that is most descriptive of the object as the front view. Sometimes this may place an object so that what is normally thought of as the front of the object is not shown in the front view of the multiview projection. For example, what is usually described as being the side of a car should be chosen as the front view, because this view is probably most descriptive and easily recognizable as a car. A view of the front of a car (grille, bumper, and windshield) is not as descriptive or as obvious as the side of the car. Furthermore, the object should be properly oriented in the front view. For instance, a car should be shown with its wheels downward in their normal operating position for the front view. The other views that are shown in addition to the front view should be views that best represent features of the object. Normally the minimum number of views necessary to accurately represent the object is used. The standard practice is to use the front, top, and right-side views. But the choice of which views to use depends on the object and which details need to be shown most clearly.

A complication that arises in multiview projections is that two different standards are used for the placement of projections. In North America (and to some extent, in Great Britain) the unfolding of the glass box approach places the top view above the front view, the right-side view to the right of the front view, and so on. This placement of views is called *third-angle projection*. But in most of the rest of the world an alternative approach for the placement of views is used. In this case the placement of views is what would result if the object were laid on the paper with its front side up for the front view and then rolled on one edge for the other views. For instance, if the object were rolled to the right so that it rests on its right side, then the left side would be facing up. So the left-side view is placed to the right of the front view. Likewise, if the object were lying on the paper with the front view up and then rolled toward the bottom of the paper, it would be resting on its bottom side, so that the top side faces upward. Thus, the top view is placed below the front view. This placement of views, known as the *first-angle projection*, simply reverses the location of the top and bottom views and the location of the left-side and right-side views with respect to the front view compared to the third-angle projection. The views themselves remain the same in both projections.

Although, the difference between the two projections is only in the placement of the views, great potential for confusion and manufacturing errors can result in engineering drawings that are used globally. To avoid misunderstanding, international projection symbols, shown in Figure 2.6, have been developed to distinguish between third-angle and first-angle projections on drawings. The symbol shows two views of a truncated cone. In the first-angle projection symbol, the truncated end of the cone (two concentric circles) is placed on the base side of the cone, as it would be in a first-angle projection. In the third-angle projection symbol, the truncated end of the cone is placed on the truncated side of the cone, as it would be in a third-angle projection. Usually these symbols appear in or near the title block of the drawing when the possibility of confusion is anticipated. Most CAD software automatically uses the third-angle projection for engineering drawings.

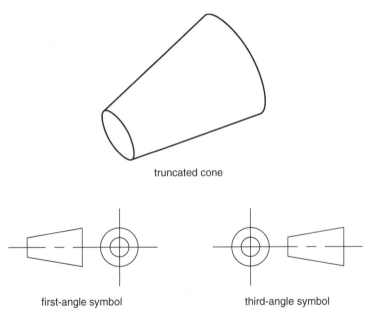

truncated cone

first-angle symbol   third-angle symbol

Figure 2.6.

A problem that frequently occurs in orthographic projections is that one of the faces of the object is at an angle to the orthographic planes that form the imaginary glass box. An example is the object shown in Figure 2.7. The circular hole with a keyway (rectangular cutout) that is perpendicular to the angled face appears in both the top view and the right-side view. However, it is distorted in both of these views, because it is on an angled plane of the object. An auxiliary view is used to avoid this distortion. In this case, a view of the object is drawn so that the angled face is parallel to the auxiliary view plane. The view is based on the viewer looking at the object along a line of sight that is perpendicular to the angled face. When viewed in this direction, the circular hole in the auxiliary view appears as a circle without any distortion. As suggested by the dash-dot lines in Figure 2.7, the auxiliary view is projected from the front view in the same way as the top and right-side views are projected. Thus, its position with respect to the front view depends on the orientation of the angled face. It is normal practice not to project hidden lines or other features that are not directly related to the angled surface.

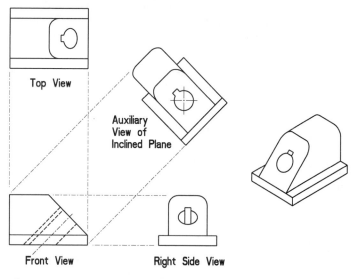

Figure 2.7.

2.4   WORKING DRAWINGS

Several types of *working drawings* are produced during the design process. Initially *freehand sketches* are used in the ideation phase of the design process. These are usually hand-drawn pictorial sketches of a concept that provide little detail, but enough visual information to convey the concept to other members of the design team. An example is the isometric sketch of a sheet metal piece that holds the blade of a pizza cutter, shown in Figure 2.8. The general shape of the object is clear, although details such as the thickness of the sheet metal and the radius of the bends in the sheet metal are not included. These conceptual sketches eventually evolve to final detailed drawings that define enough detail and information to support production.

Figure 2.8.

Detail drawings document the detailed design of individual components using orthographic views. The detail drawing is the final representation of a design that is specific enough so that all of the information necessary for the manufacture of the part is provided. As a result, it is imperative that it includes the necessary views, dimensions, and specifications required for manufacturing the part. Figure 2.9 shows an example of a detail drawing of the part of the pizza cutter that was sketched in Figure 2.8. The detail drawing includes fully dimensioned orthographic views, notation of the material that the part is to be made from, information on the acceptable tolerances for the dimensions, and a title block that records important information about the drawing. Often an isometric view is included in the detail drawing to further clarify the shape of the part. Detail drawings provide sufficient detail so that the part can be manufactured based on the drawing alone.

Figure 2.9.

Assembly drawings show how the components of a design fit together. Dimensions and other details are usually omitted in assembly drawings to enhance clarity. Several styles of assembly drawings are commonly used. Sometimes the assembly drawing is just an isometric view of the fully assembled device. But an exploded isometric view is often helpful to show the individual parts are assembled, as shown in Figure 2.10 for a pizza cutter. In some cases, a sectioned assembly, or cut-away view, shows how complicated devices are assembled. A cutting plane passes through the assembly and part of the device is removed to show the interior of the assembly. Numbers or letters can be assigned to individual parts of the assembly on the drawing and keyed to a parts list.

	QTY.	PART NO.	DESCRIPTION
1	1	52806	handle
2	1	52825	cap
3	1	42886	guard
4	1	97512	rivet
5	2	55654	arm
6	1	56483	blade

Figure 2.10.

Finally a *parts list*, or *bill of materials*, must be included with a set of working drawings. The parts list includes the part name, identification number, material, number required in the assembly, and other information (such as catalog number for standard parts such as threaded fasteners). An example is shown in Figure 2.11 for a pizza cutter. The parts list is used to ensure that all parts are ordered or manufactured and brought to the central assembly point.

	QTY.	PART NO.	DESCRIPTION
1	1	52806	handle
2	1	52825	cap
3	1	42886	guard
4	1	97512	rivet
5	2	55654	arm
6	1	56483	blade

Figure 2.11.

Taken together, the detail drawings of each individual part, the assembly drawing, and the bill of materials provide a complete set of working drawings for the manufacture of a part.

KEY TERMS

Assembly drawings
Bill of materials
Detail drawings
First-angle projection
Freehand sketches

Isometric projection
Multiview projection
Oblique projection
Orthographic projection
Perspective projection

3-D projections
Third-angle projection
Trimetric projection
Working drawings

Problems

1. Describe or sketch the front view that should be used in an orthographic projection of:

   a. a stapler.
   b. a television set.
   c. a cooking pot.
   d. a hammer.
   e. a pencil.
   f. a bicycle.
   g. an evergreen tree.
   h. a paper clip.
   i. a coffee mug.
   j. a padlock.

2. Identify the views shown in Figure 2.12 as isometric, trimetric, or perspective.

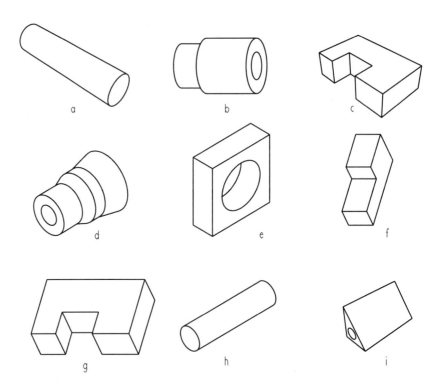

Figure 2.12.

3. For the drawings shown in Figure 2.13, determine whether the multiview projection is first-angle or third-angle.

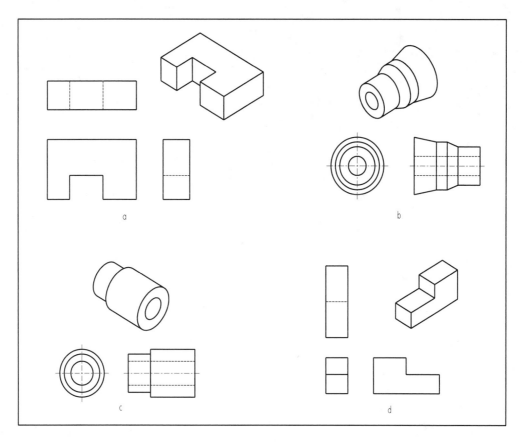

Figure 2.13.

4. Develop a bill of materials for:

    a. a pencil.
    b. a squirt gun.
    c. a click-type ball point pen.
    d. a videocassette (take an old one apart).
    e. an audio cassette (take an old one apart).
    f. a disposable camera (ask a local photo developer for a used one to take apart).
    g. eyeglasses.
    h. a household cleaner pump bottle.
    i. a claw-type staple remover.
    j. an adhesive tape dispenser.
    k. a bicycle caliper brake.
    l. a floppy disk (take an old one apart).
    m. a utility knife.
    n. a Vise-Grip wrench.

3

Freehand Sketching

OVERVIEW

In this chapter you will learn useful techniques for freehand sketching to create both two-dimensional orthographic sketches and three-dimensional pictorial sketches. You will learn how to quickly make rough sketches to convey a concept and how to make more refined sketches of objects that are more complex.

3.1   WHY FREEHAND SKETCHES?

An integral part of the creative design process is *ideation*, the generation of concepts or ideas to solve a design problem. Often freehand sketching can be used to explore and communicate mental concepts that come about in the mind's eye. The process of sketching can solidify and fill out rough concepts. Furthermore, sketching captures the ideas in a permanent form that can be used to communicate the concept to others. In this way, sketches often act as stepping stones to refine and detail the original concept or generate new ideas. Many great design ideas are first sketched on the back of an envelope or in a lab notebook, such as the freehand sketch of one of helicopter inventor Igor Sikorsky's designs, (Figure 3.1).

While computers are the workhorses for engineering graphics, initially generating ideas on a computer screen is very rare. A more common scenario is sketching an idea on paper and subsequently refining the concept on paper using more rough sketches. This often occurs simply because all

SECTIONS

- 3.1   Why Freehand Sketches?
- 3.2   Freehand Sketching Fundamentals
- 3.3   Basic Freehand Sketching
- 3.4   Advanced Freehand Sketching

OBJECTIVES

After reading this chapter, you should be able to:

- Explain why freehand sketching is important in design
- Freehand sketch lines and circles
- Sketch an oblique 3-D projection
- Sketch an isometric 3-D projection
- Sketch an orthographic multiview projection

that is needed for a freehand sketch is a pencil and a paper. Freehand sketching *quickly* translates the image of the concept in the mind's eye to paper. Engineers often communicate via rough freehand sketches to refine and improve the design. Sketches are much more useful than detailed CAD drawings early in the design process, because they are informal, quickly and easily changed, and less restrictive. It is only after clarifying the design concept by iterating through several freehand sketches that it is possible to draw the object using computer graphics. In fact, often an engineer will sit down to create a CAD drawing of an object using a freehand sketch as a guide.

This chapter focuses on the rudimentary elements of freehand technical sketching, because in many ways freehand sketching is the first step in CAD.

Figure 3.1. Helicopter inventor Igor Sikorsky's sketch of an early helicopter prototype demonstrates the visual impact of freehand sketching. (Used with permission of the Sikorsky Aircraft Corporation, Stratford, CT. ["Straight Up," by Curt Wohleber, *American Heritage of Invention and Technology,* Winter, 1993, pp. 26–39.])

3.2 FREEHAND SKETCHING FUNDAMENTALS

Freehand sketching requires few tools: just a pencil and paper. It may be tempting to use straight-edged triangles or rulers for drawing straight lines and a compass to draw circles. But these instruments often slow down the process and distract from the purpose of sketching, which is to create a quick, rough graphical representation of the image in the mind's eye. Generally sketching has three steps, although the steps are usually subconscious. First, the sketch is planned by visualizing it in the mind including the size of the sketch on the paper, the orientation of the object, and the amount of detail to be included in the sketch. Second, the sketch is outlined using very light lines to establish the orientation, proportion, and major features of the sketch. Finally, sharpening and darkening object lines and adding details develops the sketch.

All sketches are made up of a series of arcs and lines, so the ability to draw circles and straight lines is necessary. A straight line is sketched in the following way. First, sketch the endpoints of the line as dots or small crosses. Then place your pencil on the starting endpoint. Keeping your eyes on the terminal point, use a smooth continuous stroke to draw the line between the points as shown in Figure 3.2. Nearly horizontal or vertical lines are frequently easier to draw than inclined lines, so it may be helpful to shift the paper to draw the line horizontally or vertically. For long lines, it may be helpful to mark two or three points along the line and use the procedure between consecutive points or to make two or three shorter passes lightly with the pencil before a final darker line.

starting
endpoint

motion of
pencil

keep eye on
terminal endpoint

Figure 3.2.

A circle can be sketched using the following steps, illustrated in Figure 3.3a. First, draw light horizontal and vertical lines crossing at the center of the circle. Second, lightly mark the radius of the circle on each line. Finally, connect the radius marks with a curved line to form a circle. Another technique is to lightly draw a square box the same size as the circle diameter as shown in Figure 3.3b. Then lightly draw diagonals of the box and centerlines between midpoints of the sides of the box. The diagonals and centerlines should intersect at the center of the circle. Mark the radius on these lines, and sketch the circle within the box. It is sometimes helpful to mark the radius on the edge of a scrap paper and mark the radius at as many points as desired in addition to the marks on the centerlines and diagonals. Arcs are sketched in much the same way as circles, except that only a portion of the circle is sketched. It is generally easier to sketch an arc with your hand and pencil on the concave side of the arc.

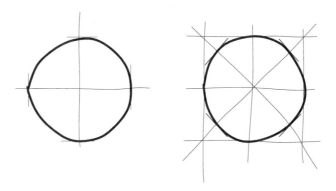

Figure 3.3.

3.3   BASIC FREEHAND SKETCHING

Many times, particularly during the conceptual stage of design, it is necessary to immediately communicate a graphical image to others. It has been said that some of the best design engineers are the ones who can sketch an idea clearly in a minute or so. The goal of the sketch in this case is not to show the details of the part, but to provide another person with a clear concept of the idea. For example, a design engineer may need to show a sketch to a manufacturing engineer to get input on the manufacturability of a part. If the concept is at an early phase, CAD drawings would not have been created yet. So the design engineer needs to use a freehand sketch of the part.

The sketch of Sikorsky's helicopter in Figure 3.1 exemplifies the power of freehand sketching. A brief glance at this sketch provides immediate insight to the concept that is being shown. One does not need to study the sketch to know what is being sketched, even if the viewer has never seen the concept before. These quick ideation sketches are not difficult to draw and require no artistic talent, just some practice.

Two types of pictorial sketches are used frequently in freehand sketching: oblique and isometric. The *oblique projection* places the principal face of the object parallel to the plane of the paper. The *isometric projection* tilts the part so that no surface of the part is in the plane of the paper. The advantage of the oblique projection is that details in the front face of the object retain their true shape. This often makes oblique freehand sketching easier than isometric sketching, where no plane is parallel to the paper. The disadvantage of the oblique projection is that it does not appear as "photorealistic" as an isometric projection. In other words, an isometric projection is similar to what a photograph of the object would look like.

3.3.1 Oblique Sketching

Often freehand sketching begins with light thin lines called *construction lines* that define enclosing boxes for the shape that is being sketched. Construction lines are used in several ways. First, the construction lines become the path for the final straight lines of the sketch. Second, the intersection of construction lines specify the length of the final lines. Third, points marked by the intersection of construction lines guide the sketching of circles and arcs. And finally, construction lines guide the proportions of the sketch. This last item is of crucial importance if the sketch is to clearly represent the object. For example, if an object is twice as wide as it is high, the proportions in the sketch must reflect this. Proper proportions of the boxes defined by the construction lines will result in proper proportions of the sketch.

An oblique freehand sketch is easy, since it begins with a two-dimensional representation of the face of the object. Figure 3.4 shows the steps in quickly sketching a part with a circular hole.

Figure 3.4.

Step 1: Horizontal and vertical construction lines are lightly drawn to outline the basic shape of the main face of the part. This is known as *blocking-in* the sketch. If you are using a pencil or felt-tip marker, press lightly when drawing the construction lines to produce a thin or light line. If you are using a ball-point pen, draw a single, light line.

Step 2: Sketch in the face of the part using the construction lines as a guide. How you sketch the outline of the part depends on the type of pen or pencil that you are using. The idea is to thicken the lines of the part compared to the construction lines. If you are using a pencil or a felt-tip marker, pressing hard for the outline of the part will result in heavy or dark lines. If you are using a ball-point pen, the line width does not depend much on how hard you press. In this case, the outline of the part is sketched with a back and forth motion of the pen to thicken the lines of the part compared to the construction lines as shown in Figure 3.4. The straight lines are usually sketched first, followed by the arcs. The circle for the hole in the part is added last to complete the face of the part.

Step 3: Sketch *receding construction lines* (lines into the plane of the paper labeled *a*) at a convenient angle. All of the receding lines must be parallel to each other and are usually at an angle of 30° to 45°. The receding lines end at the appropriate depth for the object. Then vertical and horizontal lines at the back plane of the part are added (lines labeled *b*). This blocks in the three-dimensional box enclosing the object.

Step 4: Sketch in and darken the lines outlining the part. Again it is usually easiest to sketch in the straight lines first, then the arcs, and finally any details. Because the construction lines are light compared to the outline of the part, they are not erased.

The final sketch, while rough and lacking detail, clearly shows the design intent for the part.

3.3.2 Isometric Sketching

Isometric freehand sketches are somewhat more difficult to master than oblique sketches because no face is in the plane of the paper in an isometric view. The steps to construct a simple freehand isometric sketch are shown in Figure 3.5.

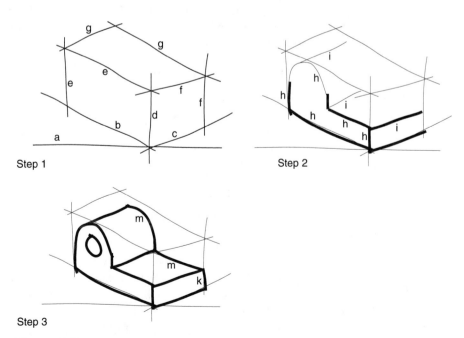

Step 1

Step 2

Step 3

Figure 3.5.

Step 1: Sketch a light horizontal line (*a*). From this line draw two intersecting lines at an angle of approximately 30° to the horizontal (*b* and *c*). Then draw a vertical line (*d*) through the intersection of the previous three lines. The three lines labeled *b*, *c*, and *d* form the isometric axes of the sketch. Next sketch the box to block in the front face of the part (*e*). These lines should be parallel to axes *b* and *d*. Similarly, sketch the lines to block in the right face (*f*) making sure that the lines are parallel to axes *c* and *d*. Finish this step by sketching lines parallel to the axes to complete the box that encloses the part (*g*).

Step 2: The outline for the front face is added by sketching in lines and curves (*h*). Then outline the front face using heavy lines. In this case, a single heavy line such as might be produced from pressing hard on a pencil or felt-tip marker is used. Next, lines are sketched to indicate the depth of the features of the front view (*i*). These lines should be parallel to axis *c*. They can be darkened after they are drawn lightly.

Step 3: Finally, a line is added to complete the back corner of the part (*k*). Lines and arcs are added to complete the back face of the part (*m*). Then the hole detail is added. Circular holes appear as ellipses in isometric views, as discussed in the next section.

The choice of whether to use an oblique projection or an isometric projection is often arbitrary. Because the oblique projection is easier to sketch, it is sometimes preferred. On the other hand, an isometric projection provides a more photorealistic image of the object.

3.4   ADVANCED FREEHAND SKETCHING

The sketching methods described in the previous section were focused on sketches in which the face of the object is in a single plane. Freehand sketching is somewhat more difficult when the face of the object is not in a single plane. The difficulty here is accurately depicting the depth of the object. Oblique and isometric projections are still useful, though somewhat more complicated than those in the previous section. In addition, orthographic projections are also valuable.

3.4.1   Freehand Oblique Sketching

An example of the steps leading to an oblique freehand sketch of a complicated object are shown in Figure 3.6. Because the face of the base of the object and the face of the upper portion of the object are in different planes, it is necessary to begin with a box that encloses the entire object before sketching either face. Some of the construction lines are removed after they are used in this example. This was done here to make the sketch more clear. However, this is not necessary in practice, if the construction lines are drawn as light lines.

Step 1: To begin, construction lines to form a box that encloses the object are drawn to block-in the sketch. Notice that the front and back faces of the box are rectangular with horizontal and vertical sides. The receding construction lines are parallel and at an angle of 30° to 45° to horizontal. The easiest way to draw this box is to first draw the front rectangle (*a*). Then draw an identical second rectangle above and to the right of the first rectangle (*b*). Finally connect the corners with receding construction lines (*c*).

Step 2: Now the front face of the base of the object can be sketched in the front rectangle. The lines are appropriately darkened.

Figure 3.6.

Step 3: Certain features of the front face of the base extend backward along or parallel to the receding construction lines. For example, the lines (*d*) forming the chamfer (angled cut on the right side of the base) can be sketched parallel to receding lines. Likewise the receding line for the upper left corner of the base can be sketched (*e*). Then the base can be finished with a horizontal line on the back face (*f*). Now it is possible to block in the upper rounded portion of the object to create a box (*g*) that encloses the upper protrusion within the larger box that encloses the entire object.

Step 4: The front face of the upper portion of the object can be sketched in this box. Then receding lines corresponding to the chamfer and the left edge of the base can be darkened. In addition, the lines forming the back face can be sketched. Note that the line forming the back edge of the chamfer is parallel to the line forming the front edge of the chamfer. Construction lines (*h*) on the front face of the upper portion are drawn to center of the circle for the hole.

Step 5: A receding construction line (*i*) extending from the peak of the front face to the plane of the back face is sketched to aid in aligning the curved outline of the back of the upper portion. The back face is identical to the front face except that it is shifted upward and to the right. This results in the left side of the back face being hidden. A darkened receding line (*k*) finishes the left side of the upper portion of the object. Finally, four construction lines (*m*) are sketched to block in the circle for the hole.

Step 6: Now the hole can be sketched in and darkened. The back edge of the hole is also added to complete the sketch. The construction lines may be erased, but usually the construction lines are retained if they are made properly as light lines.

Oblique sketching is often aided by the use of graph paper with a light, square grid. The process is identical to that shown in Figure 3.6, but it is easier to keep the proportions correct by counting the number of boxes in the grid to correspond to the approximate dimensions of the part. Graph paper further improves the sketch by helping keep lines straight as well as more accurately horizontal or vertical.

3.4.2 Isometric Sketching

Isometric freehand sketches of more complex objects start with an isometric box to block in the sketch. Then faces are sketched and additional features are blocked in. Finally details are added. The steps to construct an isometric sketch are shown in Figure 3.7 Some of the construction lines are removed after they are used, to make the sketch more clear in this figure. Normally, removing construction lines is not necessary.

Step 1: To begin, sketch a light horizontal line (*a*). From this line draw two intersecting lines at an angle of approximately 30° to the horizontal (*b* and *c*) and a vertical line (*d*) through the intersection of the previous three lines to form the isometric axes of the sketch. Finish blocking in by sketching lines (*e*) to complete the box so that it will completely enclose the object. Unlike the oblique sketch, it is often better not to sketch hidden construction lines when blocking in.

Step 2: Block in the front face of the part (*f*) so that the construction line is parallel to the isometric axis. Similarly, sketch the line to block in the right face (*g*).

Step 3: Sketch the left face and the right face and darken the lines. This completes the faces that are in the front planes of the box. Now sketch in three lines (*h*) parallel to the isometric axis (*c*). The left line (*h*) is the top edge of the base. The middle line (*h*) finishes the chamfer. The right line (*h*) is used to aid in sketching a construction line for the back edge of the base (*i*), which is sketched next.

Step 4: Now the face of the chamfer can be darkened and the angled line at the back edge of the chamfer can be added. This completes the angled face of the chamfer. Next the protrusion above the base can be blocked in with seven lines (*k*).

Step 5: The front face of the upper protrusion is sketched first using light lines. Construction lines (*m*) are added to help identify the location of the endpoints of the arc of the front and back faces of the protrusion. The rounded rear face (*n*) is sketched lightly to be identical to the front face, except that part of it is not visible. The line at the top left edge of protrusion (*o*) is added. Then all lines forming the upper portion of the object are darkened In addition, the line forming the top edge of the base on the back side is darkened.

Step 6: The details related to the hole are added next. Circles in isometric projections are difficult to draw because they appear as ellipses with their major axes at an angle to horizontal. The center of the hole is where two lines (m) intersect on the front face of the upper portion of the object. The lines (p) forming the parallelogram to enclose the ellipse for the hole are added. Each side of the parallelo-

Figure 3.7.

gram should be parallel to one of the isometric axes. The sides should be equal in length to one another.

Step 7: To help in sketching the ellipse, construction lines forming the diagonals of the parallelogram (*r*) are added.

Step 8: Now the ellipse that represents the circular hole can be sketched. A few simple points help in sketching ellipses more easily. The major axis and minor axis of the ellipse are perpendicular to one another. The major and minor axes also coincide with the diagonals of the parallelogram enclosing the ellipse (*r*). The ellipse touches the parallelogram at the midpoints of the sides of the parallelogram. Start drawing the hole by sketching short elliptical arcs between the midpoints of the parallelogram on either side of the minor axis. Finish the hole by sketching sharply curved elliptical arcs between the midpoints of the parallelogram on either side of the major axis of the hole. Finally, darken and make heavy the lines outlining the hole and any remaining edges of the part.

Isometric sketching is made substantially easier by the use of isometric grid paper. This paper has a grid of lines at horizontal and 30° to horizontal (corresponding to lines b, c, and d in Figure 3.7). The procedure for using isometric grid paper is the same as that described above, but using the isometric grid paper keeps proportions of the part consistent. One simply counts grid boxes to approximate the dimensions of the object. The grid paper also aids in sketching straight lines parallel to the isometric axes.

3.4.3 Orthographic Sketching

In some cases it is necessary to sketch orthographic projection views rather than oblique or isometric pictorial views. Because orthographic views are two-dimensional representations, they are not as difficult to sketch as pictorial views. But there are several techniques that make freehand sketching of orthographic views easier and more efficient. The process for sketching three orthographic views of the object in the previous two figures is shown in Figure 3.8.

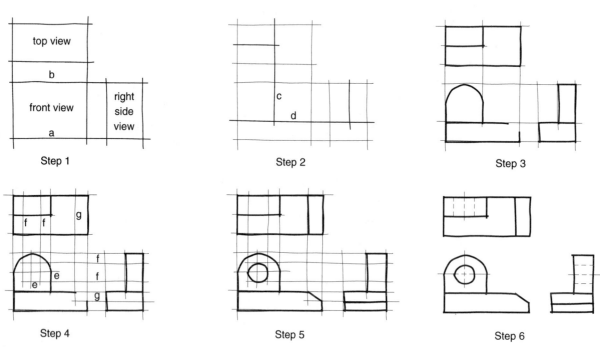

Figure 3.8.

Step 1: Begin by blocking in the front, top, and side views of the object using the overall width, height, and depth. The construction lines extend between views to properly align the views and maintain the same dimension in different views. For instance, line (*a*) represents the bottom edge and line (*b*) represents the top edge in both the front view and the right-side view. The distance between lines (*a*) and (*b*) is the height dimension in both views. The space between the views should be large enough so that the drawing does not look crowded and should be the same between all views.

Step 2: In the second step the upper protrusion is blocked in. Note that line (*c*) extends across the top and front views, to assure that the width of the protrusion is consistent in both views. Likewise, line (*d*) extends across the front and right-side views.

Step 3: The outline of the object is darkened to clearly show the shape of the object in all three views. Care must be taken in darkening lines. For instance, the right corner of the front view should not be darkened, because the detail of the chamfer has not yet been added.

Step 4: Construction lines for the holes and other details are added next. The center of the hole is positioned with construction lines (*e*). Then construction lines that block in the hole (*f*) are drawn. These construction lines extend between views to project the hole to the top view and to the right-side view. Construction lines extending between views (*g*) are also added for the chamfer.

Step 5: Now the hole and chamfer are sketched and darkened to show the completed object.

Step 6: Finally, centerlines (long-dash, short-dash) that indicate the center of the hole are added. Hidden lines (dashed lines) that indicate lines hidden behind a surface are also added. Construction lines may be erased as was done in this figure, but this is not usually necessary.

The quality of the sketch can often be improved by using square grid graph paper to keep proportions and act as a guide for horizontal and vertical lines. Some engineers prefer to use a straight-edge to produce a nicer sketch, but this is usually not necessary with practice and sufficient care in sketching.

PROFESSIONAL SUCCESS

Will Freehand Sketching Ever Become Obsolete?

CAD has almost totally eliminated pencil and paper drawings. But what about pencil and paper freehand sketching? Although many computers and palm-tops offer sketching programs, it is unlikely that freehand sketching will disappear soon. Just as it is easier to do a calculation in your head or on a piece of scratch paper rather than finding a calculator and punching in the numbers, it is easier to sketch an image on a piece of paper (or a napkin!) than to find a computer, log in, and start the appropriate "paint" program. Pencil and paper freehand sketches are quick, effi-cient, easily modified, and easily conveyed to others. And all that is needed is a pencil and a scrap of paper.

Even if the pencil and paper are totally replaced by palm-top computers some day, freehand sketching skills will still be useful. Instead of using a pencil on a piece of paper, we will use a stylus on a touch screen. The only difference is the medium. The freehand sketching techniques themselves are unlikely to change much. Perhaps one could imagine sketching software that assists in generating oblique or isometric sketches as they are drawn at some time in the future.

KEY TERMS

Blocking-in	Ideation
Construction lines	Receding construction lines

Problems

For all of the problems, the items shown in Figure 3.9, 3.10, and 3.11 are 2 inches wide, 1.5 inches high, and 1 inch deep. The holes in Figure 3.9b, 3.9c, 3.9d, 3.10c, and 3.10d are through holes. The hole in Figure 3.10b is through the front face only.

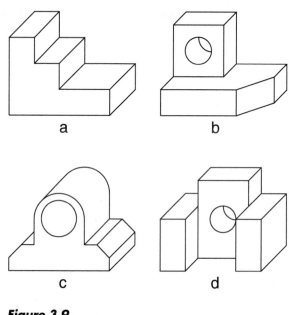

a   b

c   d

Figure 3.9.

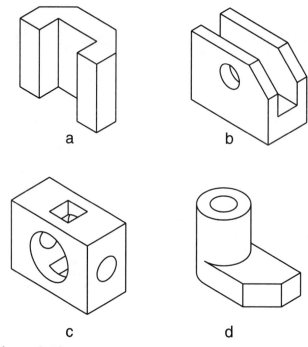

a   b

c   d

Figure 3.10.

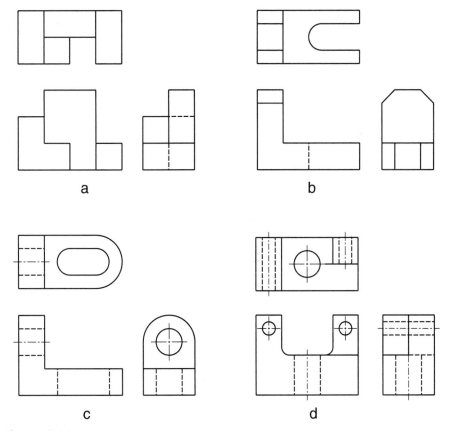

Figure 3.11.

1. Create freehand oblique sketches of the objects in Figure 3.9. (The objects are shown as oblique projections, so you must simply recreate the drawing by freehand sketching.)

2. Create freehand oblique sketches of the objects in Figure 3.10. (The objects are shown as isometric projections.)

3. Create freehand oblique sketches of the objects in Figure 3.11. (The objects are shown as orthographic projections.)

4. Create freehand isometric sketches of the objects in Figure 3.10. (The objects are shown as isometric projections, so you must simply recreate the drawing by freehand sketching.)

5. Create freehand isometric sketches of the objects in Figure 3.9. (The objects are shown as oblique projections.)

6. Create freehand isometric sketches of the objects in Figure 3.11. (The objects are shown as orthographic projections.)

7. Create freehand orthographic sketches of the objects in Figure 3.11. (The objects are shown as orthographic projections, so you must simply recreate the drawing by freehand sketching.)

8. Create freehand orthographic sketches of the objects in Figure 3.9. (The objects are shown as oblique projections.)

9. Create freehand orthographic sketches of the objects in Figure 3.10. (The objects are shown as isometric projections.)

4

Computer Aided Design and Drafting

OVERVIEW

Several different models for representing a part's geometry have been used in CAD, including 2-D models and 3-D wireframe and surface models. However, solids modeling is the current state-of-the-art in CAD. Solids modeling has the inherent advantage of more accurately representing the "design intent" of the part that modeled.

4.1   CAD MODELS

A CAD *model* is a computer representation of an object or part. It can be thought of as a "virtual" part in that it exists only as a computer image. The model is an engineering document of record. It contains all of the design information including geometry, dimensions, tolerances, materials, and manufacturing information. A CAD model replaces the paper blueprints and engineering drawings of just a few decades ago.

The simplest model used in CAD is a *2-D model*. This model is essentially the computer graphics equivalent to an orthographic projection created using a pencil and paper as shown in Figure 4.1. A 2-D model represents a three-dimensional object with several views, each showing the view of one side of the object projected onto a plane. On the computer, a 2-D model appears as lines and curves on a flat surface, and it is up to the engineer or designer to interpret several views to create a mind's eye image of the three-dimensional object. The CAD software electronically stores

SECTIONS

- 4.1   CAD Models
- 4.2   CAD and Solids Modeling
- 4.3   The Nature of Solids Modeling

OBJECTIVES

After reading this chapter, you should be able to:

- Differentiate 2-D, 2 1/2-D, and 3-D models
- Differentiate wireframe, surface, and solids models
- Explain how a cross-section is extruded or revolved to create a solids model
- Explain the analogy between creating a solids model of a part and machining the part
- Explain feature-based modeling
- Explain constraint-based modeling
- Explain history-based modeling

the 2-D model as several views but does not inherently "know" that the views can be connected to one another to form a three-dimensional object.

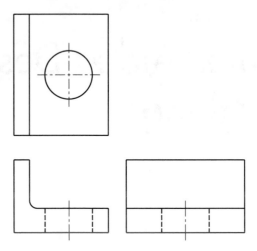

Figure 4.1.

A *2 1/2-D model* has a third dimension that is recognized by the CAD software, but the third dimension is simply an extrusion of a two-dimensional shape. In this case, the CAD software electronically stores the cross-section shape of the object and the depth that it is *extruded*, or stretched, in the third dimension. Thus, the only objects that can be represented using a 2 1/2-D model model are those that have a constant cross-section. For instance, the simple shape shown in Figure 4.2 is the 2 1/2-D model of the object shown using a 2-D model in Figure 4.1 without the hole. The object, including the hole, cannot be represented as a 2 1/2-D model because the hole results in a change in the cross-section. A 2 1/2-D model could be used to represent a shape such as an I-beam or C-channel.

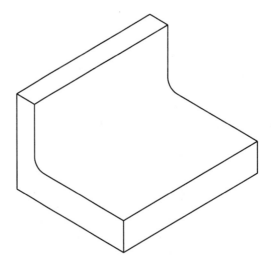

Figure 4.2.

A *3-D model* is the most general model used in CAD software. In this case, the CAD software electronically stores the entire three-dimensional shape of the object. Most simply, one can think of the software storing the three-dimensional coordinates of various points on the object and then defining how these points are connected. Of course, a two-dimensional drawing representing the orthographic views of the object can be automatically created from the 3-D model. The current generation of CAD software is based on 3-D models.

The simplest 3-D model is a *wireframe model*. Figure 4.3a shows a wireframe representation of the object that we have been considering. In a wireframe model, only edges of the object are represented. Thus, the CAD software needs to store the locations of the vertices (intersections of lines) and which vertices are to be connected to each other by lines. A circle or curved section can be represented by a series of closely spaced vertices with linear edges connecting them or as an arc defined by a center, radius, and end points. Before more powerful computers were available, the low storage requirements of a wireframe model made it quite popular. The problem inherent in a wireframe representation is quite clear in Figure 4.3a—it is difficult to interpret the drawing because all of the edges are visible. In fact, there are many situations where a wireframe representation cannot be unambiguously interpreted because it is not clear which lines would remain hidden if the surfaces of the part were filled in. This problem can be solved using various methods to remove "hidden lines" from the wireframe. This is tricky because the CAD software has to first determine how the edges of the wireframe form a surface and then determine if that surface hides any lines. Figure 4.3b shows the object with the hidden lines shown as dashed lines to indicate that they are behind a surface. 3-D models other than wireframes are able to deal with hidden lines more effectively and offer other advantages.

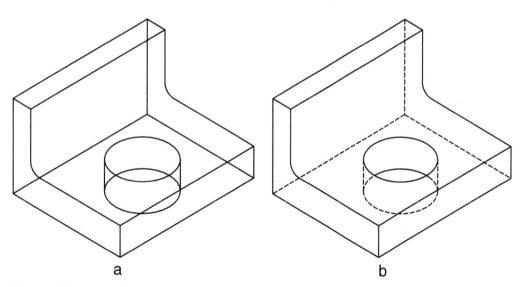

a   b

Figure 4.3.

A 3-D *surface model* defines the object in terms of surfaces such as plates (flat) and shells (curved) in addition to edges. This makes it easy to determine whether a line or surface is hidden, because the surfaces are defined. Using a surface model also permits the construction of smoothly curved surfaces, such as a circular hole, rather than using many line segments to form a curved line as with some wireframe models. It is also possible to "fair" or smooth one surface into another surface to provide a sculpted shape.

Figure 4.4a shows the simple object with hidden lines removed altogether. Although a surface model can be displayed to look like a wireframe, either with or without hidden lines, it can also be displayed by assigning different degrees of shading to the surfaces. A virtual light source is assumed to be near the object to provide a three-dimensional lighting effect as shown in Figure 4.4b. This is known as *shading* or *rendering*. Although a variety of shading techniques are available, all depend on determining how much light from a virtual light source strikes each portion of the virtual object. The use of shading provides a very realistic image of the object, which permits a much more vivid communication of the nature of the object. However, the problem that remains with a surface model is that the interior of the object remains undefined. The surface model just represents the shell of the object.

a   b

Figure 4.4.

4.2   CAD AND SOLIDS MODELING

Solids modeling, the current state-of-the-art in CAD, is the most sophisticated method of representing an object. Unlike wireframe or surface models, a solids model represents an object in the virtual environment just as it exists in reality, having volume as well as surfaces and edges. In this way, the interior of the object is represented in the model as well as the outer surfaces.

The first attempt at solids modeling was a technique known as *constructive solid geometry* or *primitive modeling*, which is based on the combination of geometric primitives such as right rectangular prisms (blocks), right triangular prisms (wedges), spheres, cones, and cylinders. Each primitive could be scaled to the desired size, translated to the desired position, and rotated to the desired orientation. Then one primitive could be added to or subtracted from other primitives to make up a complex object. It is easy to see how two blocks plus a negative cylinder (hole), as shown in Figure 4.5, could represent an object similar to the one shown in Figure 4.4. (The rounded surface, or fillet, at the corner between the horizontal and vertical surfaces has been omitted.) The problem with constructive solid geometry was that the mental process of creating a solid model based on geometric primitives was much more abstract than the mental processes required for designing real world objects.

Figure 4.5.

Constraint-based solids modeling overcomes the weakness of constructive solid geometry modeling by making the modeling process more intuitive. Instead of piecing together geometric primitives, the constraint-based modeling process begins with the creation of a 2-D sketch of the profile for the cross-section of the part. Here, "sketch" is the operative word. The sketch of the cross-section begins much like the freehand sketch of the face of an object in an oblique view. The only difference is that CAD software draws straight lines and perfect arcs. The initial sketch need not be particularly accurate; it need only reflect the basic geometry of the part's cross-sectional shape. Details of the cross-section are added later. The next step is to constrain the 2-D sketch by adding enough dimensions and parameters to completely define the shape and size of the 2-D profile. The name *constraint-based modeling* arises because the shape of the initial 2-D sketch is "constrained" by adding dimensions to the sketch. Finally, a three-dimensional object is created by *revolving* or extruding the 2-D sketched profile. Figure 4.6 shows the result of revolving a simple L-shaped cross-section by 270° about an axis and extruding the same L-shaped cross-section along an axis. In either case, these solid bodies form the basic geometric solid shapes of the part. Other features can be subsequently added to modify the basic solid shape.

Figure 4.6.

Once the solids model is generated, all of the surfaces are automatically defined, so it is possible to shade it in the same way as a surface model is shaded. It is also easy to generate 2-D orthographic views of the object. This is a major advancement from other modeling schemes. With solids modeling, the two-dimensional drawings are produced as views of the three-dimensional virtual model of the object. In traditional CAD, the three-dimensional model is derived from the two-dimensional drawings. One might look at solids modeling as the sculpting of a virtual solid volume of material. Because the volume of the object is properly represented in a solids model, it is possible to slice through the object and show a view of the object that displays the interior detail. Once several solid objects have been created, they can be assembled in a virtual environment to make sure that they fit together and to visualize the assembled product.

PROFESSIONAL SUCCESS

How Solids Models are Used

Solids models are useful for purposes other than visualization. The solids model contains a complete mathematical representation of the object, inside and out. This mathematical representation is converted easily into specialized computer code that can be used for stress analysis, heat transfer analysis, fluid flow analysis, and computer-aided manufacturing.

Finite Element Analysis (FEA) is a method to subdivide the object under study into many small, simply shaped elements. The simple geometry of the finite elements allows the relatively simple application of the appropriate equations for the stress and deformation of the elements or for the heat transfer between elements. By properly relating nearby elements to one another, all of these equations for each element can be solved simultaneously. The final result is the stress, deformation (strain), or temperature throughout the object for a given applied load or thermal condition. The stress, deformation, or temperature is often displayed as a color map on the solid model, visually displaying the regions of high stress or temperature. From these results, the engineer can redesign the object to avoid stress concentrations, large deformations, or undesirable temperatures. Likewise, the fluid flow through or around an object can be calculated by applying similar FEA tech-

niques using the equations of fluid flow to the regions bounded by the object. These analysis tools have greatly improved and enhanced the design of many products.

Solids models can also be used in the manufacturing process. Software is available to automatically generate machine tool paths to machine the object based on the solids model. Then the software simulates the removal of material from an initial block of material on the computer. This allows the engineer to optimize the machining operation. These tool paths can be downloaded onto Computer Numerical Control (CNC) machine tools. The CNC machine tool automatically removes the desired surfaces from a block of material with high precision, allowing many identical parts to be machined automatically based directly on the solids model. Alternatively, the solids model can be downloaded onto a rapid prototyping machine. This machine automatically drives lasers that solidify liquid plastic resin, drives a robotic system to dispense a fine bead of molten plastic, or drives a laser that cuts layers of paper. In all three cases, layers are built up to become a physical model of the computer solids model, often within just a few hours. These manufacturing tools have made it possible to obtain parts from CAD models in several hours instead of several weeks.

4.3 THE NATURE OF SOLIDS MODELING

Solids modeling grew steadily in the 1980s, but it was not until Pro/ENGINEER® was introduced in 1988 and SolidWorks® was introduced in the 1990s that solids modeling delivered on its promised gains in productivity. These gains result from four characteristics of the software: feature-based, constraint-based, and parametric, associative modeling.

Feature-based modeling attempts to make the modeling process more efficient by creating and modifying geometric *features* of a solid model in a way that represents how geometries are created using common manufacturing processes. Features in a part have a direct analogy to geometries that can be manufactured or machined. A *base feature* is a solid model that is roughly the size and shape of the part that is to be modeled. The base feature is the 3-D solid created by revolving or extruding a cross-section, such as those shown in Figure 4.6. It can be thought of as the initial work block. All subsequent features reference the base feature either directly or indirectly. Additional features shape or refine the base feature. Examples of additional features include holes or cuts in the initial work block.

The analogy between feature-based modeling and common manufacturing processes is demonstrated in Figure 4.7 for making a handle for a pizza cutter. Beginning at the top of the figure we follow the steps that an engineer would use to create a solid model, or virtual part, on the left and the steps that a machinist would take to create the same physical part in a machine shop on the right. The engineer using solids modeling software begins by creating a two-dimensional profile, or *cross-section*, of a part, in this case a circle (shown in isometric projection). The analogous step by a machinist is to choose a circular bar stock of material with the correct diameter. Next the engineer *extrudes*, or stretches, the circular cross-section along the axis perpendicular to the plane of the circle to create a three-dimensional base feature (a cylinder in this case). The equivalent action by a machinist is to cut off a length of bar stock to create an initial work block. Now the engineer adds features by cutting away material on the left end to reduce the diameter and by rounding the right end of the cylinder. The machinist performs similar operations on a lathe to remove material from the cylinder. Next the engineer creates a circular cut to form a hole through the cylinder on the right end. The machinist drills a hole in the right end of the cylinder. Finally, the engineer creates a pattern of groove cuts around the handle. Likewise the machinist cuts a series of grooves using a lathe. In similar fashion a geometric shape could be added to the base feature in the solids model, analogous to a machinist welding a piece of metal to the work block. Feature-based techniques give the engineer the ability to easily create and modify common manufactured features. As a result, planning the manufacture of a part is facilitated by the correspondence between the features and the processes required to make them.

Constraint-based modeling permits the engineer or designer to incorporate "intelligence" into the design. Often this is referred to as *design intent*. Unlike traditional CAD software, the initial sketch of a two-dimensional profile in constraint-based solids modeling does not need to be created with a great deal of accuracy. It just needs to represent the basic geometry of the cross-section. The exact size and shape of the profile is defined through assigning enough parameters to fully "constrain" it. Some of this happens automatically. For example, if two nearly parallel edges are within some preset tolerance range of parallel (say 5 degrees), then the edges are automatically constrained to be parallel. As the part is resized, these edges will always remain parallel no matter what other changes are made. Likewise, if a hole is constrained to be at a certain distance from an edge, it will automatically remain at that distance from the edge, even if the edge is moved. This differs from traditional CAD where both the hole and the edge would each be fixed at a particular coordinate location. If the edge were moved, the hole location would need to be respecified so that the hole will remain the same distance from the edge. The advantage of constraint-based modeling is that the design intent of the engineer remains intact as the part is modified.

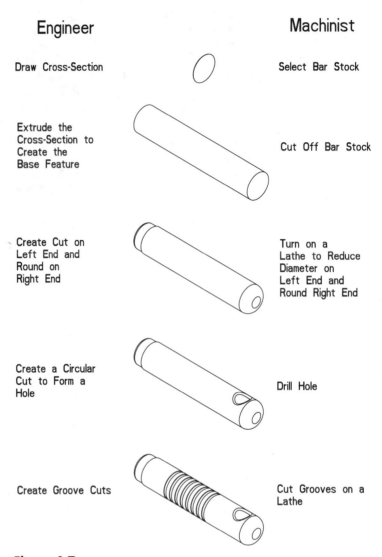

Engineer   Machinist

Draw Cross-Section — Select Bar Stock

Extrude the Cross-Section to Create the Base Feature — Cut Off Bar Stock

Create Cut on Left End and Round on Right End — Turn on a Lathe to Reduce Diameter on Left End and Round Right End

Create a Circular Cut to Form a Hole — Drill Hole

Create Groove Cuts — Cut Grooves on a Lathe

Figure 4.7.

Another aspect of constraint-based modeling is that the model is *parametric*. This means that parameters of the model may be modified to change the geometry of the model. A dimension is a simple example of a parameter. When a dimension is changed, the geometry of the part is updated. Thus, the *parameter drives the geometry*. This is in contrast to other modeling systems in which the geometry is changed, say by stretching a part, and the dimension updates itself to reflect the stretched part. An additional feature of parametric modeling is that parameters can reference other parameters through relations or equations. For example, the position of the hole in Figure 4.4 could be specified with numerical values, say 40 mm from the right side of the part. Or the position of the hole could be specified parametrically, so that the center of the hole is located at a position that is one-half of the total length of the part. If the total length was specified as 80 mm, then the hole would be 40 mm from either side. However, if the length was changed to 100 mm, then the hole would automatically be positioned at half of this distance, or 50 mm from either side. Thus, no matter what the length of the part is, the hole stays in the middle of its length. The power of this approach is that when one dimension is modified, all linked dimensions are updated according to specified mathematical relations, instead of having to update all related dimensions individually.

The last aspect of constraint-based modeling is that the order in which parts are created is critical. This is known as *history-based modeling*. For example, a hole should not be created before a solid volume of material in which the hole occurs has been modeled. If the solid volume is deleted, then the hole should be deleted with it. This is known as a *parent-child relation*. The child (hole) cannot exist without the parent (solid volume) existing first. Parent-child relations are critical to maintaining design intent in a part. Most solids modeling software recognizes that if you delete a feature with a hole in it, you do not want the hole to remain floating around without being attached to a feature. Consequently, careful thought and planning of the base feature and initial additional features can have a significant effect on the ease of adding subsequent features and making modifications.

The *associative* character of solids modeling software causes modifications in one object to "ripple through" all associated objects. For instance, suppose that you change the diameter of a hole on the engineering drawing that was created based on your original solid model. The diameter of the hole will be automatically changed in the solid model of the part, too. In addition, the diameter of the hole will be updated on any assembly that includes that part. Similarly, changing the dimension in the part model will automatically result in updated values of that dimension in the drawing or assembly incorporating the part. This aspect of solids modeling software makes the modification of parts much easier and less prone to error.

As a result of being feature-based, constraint-based, and associative, solids modeling captures "design intent," not just the design. This comes about because solids modeling software incorporates engineering knowledge into the solid model with features, constraints, and relationships that preserve the intended geometrical relationships in the model.

PROFESSIONAL SUCCESS

Has CAD Impacted Design?

CAD has not only automated and provided a more accurate means of creating engineering drawings, it has created a new paradigm for design. Parts can be modeled, visualized, revised, and improved on the computer screen before any engineering drawings have even been created. Parts that have been modeled can be assembled in the virtual environment of the computer. The relative motion of moving parts can be animated on the computer. The stresses in the parts can be assessed computationally and the part redesigned to minimize stress concentrations. The flow of fluids through the part or around the part can be modeled computationally. The machine tool path or mold-filling flow to fabricate the part can be modeled on the computer. The part model can be downloaded to a rapid prototyping system that can create a physical model of the part in a few hours with virtually no human intervention.

Although some of the analyses described above are outside of the capability of the CAD software itself, the ability to produce a CAD model makes these analyses possible. Without the ability to create a complex solid geometry computationally using CAD, much of the engineering analysis that is used to design, improve, and evaluate a new design would be much more difficult.

KEY TERMS

Associative modeling	Feature	Revolve
Base Feature	Feature-based modeling	Solids modeling
Constraint-based modeling	History-based modeling	Surface model
Design intent	Parametric modeling	Wireframe model
Extrude	Parent-child relation	

Problems

1. The wireframe model shown in Figure 4.8 is a small cube centered within a large cube. The nearest corners of the two cubes are connected. For instance, the lower left front corners of both cubes are connected with a line. This wireframe model is ambiguous; that is, it could represent several different solid bodies. Freehand sketch or trace the figure. Then use dashed lines to indicate hidden lines to show two different solids model configurations that are possible for this wireframe model.

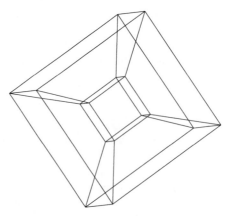

Figure 4.8.

2. Using a technique similar to that shown in Figure 4.5, freehand sketch the geometric primitives that could be added or subtracted to form the objects shown in Figure 4.9. (The hole in Figure 4.9b is only through the face shown, not the back face. The holes in Figures 4.9c and 4.9d are through holes.)

a   b

c   d

Figure 4.9.

3. Sketch the cross-section that should be revolved to create the objects shown in Figure 4.10. Include the axis about which the cross-section is revolved in the sketch.

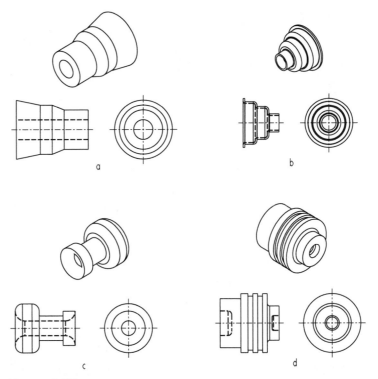

Figure 4.10.

4. Sketch at least two cross-sections that could be extruded to form a base feature for the objects shown in Figure 4.11. Use cross-sections that can be extruded into a base feature from which material is only removed, not added. Make sure that the cross-sections that you propose minimize the number of additional features necessary to modify the base feature. (The hole in Figure 4.11a is only through the face shown, not the back face. The holes in Figure 4.11b, 4.11c, and 4.11e are through holes.)

5. Consider the following objects. Sketch what the base feature would look like. List what features would be added to model the object. The type of base feature to be used (extrude, revolve, or both) is noted.

    a. hexagonal cross-section wooden pencil that is sharpened to a point (do not include eraser)—extrude.

    b. plastic 35 mm film container and cap—extrude and revolve.

    c. nail—extrude and revolve.

    d. push pin—revolve.

    e. baseball bat—revolve.

    f. broom handle—revolve and extrude.

    g. ceiling fan blade—extrude.

Figure 4.11.

> h.   cinder block—extrude.
> i.   compact disk—revolve and extrude.
> j.   single staple (before being deformed)—extrude.
> k.   automobile tire—revolve.
> l.   gear—extrude.
> m.   round toothpick—revolve.

6.   Describe parametric modeling and give an example of a family of parts where parametric modeling would be useful.

7.   Describe constraint-based modeling and explain how it relates to design intent.

5

Standard Practice for Engineering Drawings

OVERVIEW

Drawing standards and conventions are used to clarify engineering drawings and simplify their creation. For example, standard sizes for drawings are used, and standard types and weights of lines designate different items on a drawing. The proper placement of dimensions on drawings is helpful in making the drawing more readable. In addition, internal or small details of the part can be displayed using special views. Screw threads are difficult to draw, so standard representations are used for screw threads.

5.1 INTRODUCTION TO DRAWING STANDARDS

Not only is the accurate and clear depiction of details of a part necessary in an engineering drawing, but it is also necessary that the drawing conforms to commonly accepted standards, or conventions. There are two reasons for the existence of standards and conventions. First, using standard symbols and projections ensures clear interpretation of the drawing by the viewer. An example of the problem of differing presentations is the use of the third-angle orthographic projection in North America and the first-angle orthographic projection elsewhere in the world. (Recall that the top and bottom views and the right and left views are reversed in third-angle and first-angle projections.) This can bring about confusion if, for example, a U. S. engineer tries

SECTIONS

- 5.1 Introduction to Drawing Standards
- 5.2 Sheet Layouts
- 5.3 Lines
- 5.4 Dimension Placement and Conventions
- 5.5 Section and Detail Views
- 5.6 Fasteners and Screw Threads
- 5.7 Assembly Drawings

OBJECTIVES

After reading this chapter, you should be able to:

- Explain why drawing standards are used
- Choose the proper sheet layout
- Read and understand the scale of a drawing
- Differentiate different linetypes used in engineering drawings
- Properly place dimensions on a drawing
- Read and understand section and detail views on a drawing
- Read and understand screw thread designations on a drawing
- Read an assembly drawing

to interpret a German drawing. If a single standard existed in the world, there would be no confusion. A second reason for using conventions is to simplify the task of creating engineering drawings. For example, the symbol Ø associated with a dimension indicates that the dimension is a diameter of a circular feature. Without this convention, it would be more difficult to unambiguously represent diameter dimensions on a drawing.

Most CAD programs automatically use a standardized presentation of drawings, usually based on standards from the American National Standards Institute (ANSI) in the United States and the International Standards Organization (ISO) in the rest of the world. Even with the use of CAD software to assure standard presentation in drawings, it is up to the engineer to implement drawing standards in some cases. An example is in the placement of dimensions in a drawing. The user controls where dimensions appear on the drawing. Even in cases where the CAD software automatically implements drawing standards, it is still necessary to understand what the conventions mean so they are properly interpreted.

5.2   SHEET LAYOUTS

Engineering drawings are created on standard size sheets that are designated by the code indicated in Table 5.1. The ISO drawing sizes are just slightly smaller than ANSI sizes. Engineering drawings are almost always done in "landscape" orientation so that the long side of the drawing is horizontal. The choice of drawing size depends upon the complexity of the object depicted in the drawing. The drawing should be sized so that the projections of the part, dimensions, and notes all fall within the borders of the drawing with adequate spacing so that the drawing is not cluttered. The drawing should be large enough so that all details are readily evident and readable. As a result, regardless of the physical size of the part, simple parts are usually drawn on smaller sheets because it is not necessary to show much detail. Complex parts are usually drawn on larger sheets to readily show adequate detail. When using CAD software, it is necessary to consider the printed drawing size in addition to the drawing size on the computer screen. On the computer screen, it is possible to Zoom In or Zoom Out to read detail, but this cannot be done for a printed drawing. Consequently, the sheet size should be chosen carefully.

TABLE 5-1   Standard Sheet Sizes

ANSI	ISO
A—8.50″ × 11.00″	A4—210 mm × 297 mm
B—11.00″ × 17.00″	A3—297 mm × 420 mm
C—17.00″ × 22.00″	A2—420 mm × 594 mm
D—22.00″ × 34.00″	A1—594 mm × 841 mm
E—34.00″ × 44.00″	A0—841 mm × 1189 mm

The *title block* on a drawing records important information about the working drawing. Normally the title block of a drawing is in the lower right corner of the drawing. ANSI standard title blocks can be used, but often individual companies use their own standard title block. The title block, such as the one shown in Figure 5.1, is used primarily for drawing control within a company. The title block includes information regarding the part depicted in the drawing, such as its name and part number; the person who created the drawing; persons who checked or approved the drawing; dates the drawing was created, revised, checked, and approved; the drawing number; and the name of the company. This information allows the company to track the drawing within the company if questions arise about the design.

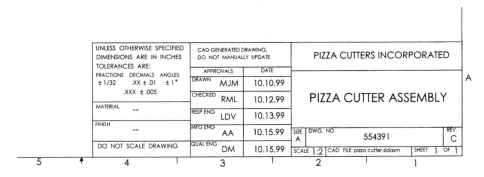

Figure 5.1.

Other information can appear in the title block to provide details about the manufacture of the part, such as the material for the part, the general tolerances, the surface finish specifications, and general instructions about manufacturing the part (such as "remove all burrs"). Many times, though, this information is shown as notes on the drawing rather than in the title block.

Finally the title block provides information about the drawing itself, such as the units used in dimensioning (typically inches or millimeters) and the scale of the drawing. The *scale* of the drawing indicates the ratio of the size on the drawing to the size of the actual object. It can be thought of as the degree to which the object is enlarged or shrunk in the drawing. Scales on drawings are denoted in several ways. For instance, consider an object that is represented in the drawing as half of its actual size. Then the drawing is created such that 1 inch on the drawing represents 2 inches on the physical object. This can be reported in the Scale box in the title block as HALF SCALE, HALF SIZE, 1 = 2, 1:2, or 1/2″ = 1″. In the numerical representations, the left number is the length on the drawing and the right number is the length of the actual object. A part that is drawn twice as large as its physical size would be denoted as DOUBLE, 2× (denoting "2 times"), 2 = 1, or 2:1. In some cases, such as architectural drawings, the scale can be quite large. For instance 1/4″ = 1′-0″ corresponds to a scale of 1:48, where ″ denotes inches and ′ denotes feet. Common scales used in engineering drawing are 1 = 1, 1 = 2, 1 = 4, 1 = 8, and 1 = 10 for English units and 1:1, 1:2, 1:5, 1:10, 1:20, 1:50, and 1:100 for metric units. A 1:1 scale is often specified as FULL SCALE or FULL SIZE.

5.3   LINES

The types of lines used in engineering drawing are standardized. Different linetypes have different meanings and a knowledge of linetypes makes interpreting drawings easier. The linetypes vary in thickness and in the length and number of dashes in dashed lines. Most CAD software automatically produces linetypes that correspond correctly to the application. Figure 5.2 shows some of the more commonly used linetypes, each of which is described below:

> *Visible lines* (thick, solid lines) represent the outline of the object that can be seen in the current view.
>
> *Hidden lines* (thin, dashed lines) represent features that are hidden behind surfaces in the current view.

Centerlines or *symmetry lines* (thin, long-dash/short-dash lines) mark axes of rotationally symmetric parts or features.

Dimension lines (thin lines with arrowheads at each end) indicate sizes in the drawing.

Extension lines or *witness lines* (thin lines) extend from the object to the dimension line to indicate which feature is associated with the dimension.

Leader lines (thin, solid lines terminated with arrowheads) are used to indicate a feature with which a dimension or note is associated.

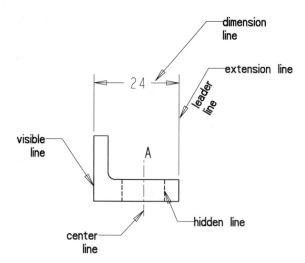

Figure 5.2.

Other linetypes are used for break lines (lines with zigzags) that show where an object is "broken" in the drawing to save drawing space or reveal interior features and cutting plane lines (thick lines with double short dashes and perpendicular arrows at each end) that show the location of cutting planes for section views.

The arrowheads for dimension lines, leader lines, and other types of lines may be filled or not filled. Most CAD software uses unfilled arrowheads as a default, although many engineering graphics textbooks prefer filled arrowheads.

It is common in technical drawings that two lines in a particular view coincide. When this happens, a convention called the *precedence of lines* dictates which line is shown. Visible lines have precedence over hidden lines, which have precedence over centerlines, which have precedence over extension lines.

Conventions also exist for the intersection of lines in a drawing. For instance, an extension line is always drawn so that there is a slight, visible gap between its end and the outline of the object as shown in Figure 5.2. Similarly, perpendicular centerlines that cross are drawn so that the short dashes of each line intersect to form a small cross. Fortunately, most CAD software takes care of all of the details related to linetypes automatically. The engineer or designer need only know how to properly interpret the various linetypes.

5.4   DIMENSION PLACEMENT AND CONVENTIONS

Engineering drawings are usually drawn to scale, but it is still necessary to specify numerical dimensions for convenience and to ensure accuracy. In general, sufficient dimensions must be provided to define precisely the geometry of the part, but redundant dimensions should be avoided. Furthermore, dimensions on the drawing should reflect the way the part is made or the critical dimensions of the part. For example, consider the three drawings of a plate with two holes shown in Figure 5.3. The left drawing is over-dimensioned, having one redundant dimension. It is clear that given the distances from the left edge to the left hole, the distance from the right edge to the right hole, and the overall width of the plate, the distance between the holes has to be 6-mm. Thus, the 6-mm dimension could be omitted, suggesting that the distance between each hole and the nearest edge of the plate is most important. But suppose that the distance between the holes is critical. This might be necessary if the holes in the plate are supposed to align with another part. Then the dimensions on the middle drawing would clearly show that the distance between the holes is most important by omitting the dimension from the right hole to the right edge. The right drawing indicates that the distance from the left edge to each hole, not the distance between the holes, is critical. The engineer or designer must consider the optimal way to display dimensions to clearly show which dimensions are most important.

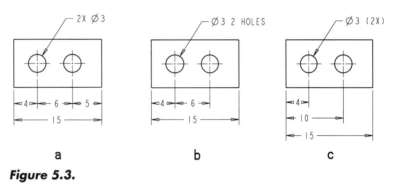

a   b   c

Figure 5.3.

It is natural for a person reading the drawings to look for the dimensions of a feature in the view where the feature occurs in its most characteristic shape and where it is visible (as opposed to hidden). This is known as *contour dimensioning*. For example, the location and size of a hole should be dimensioned in the view where the hole appears as a circle. Likewise, it is best to show the overall dimensions of the object in a view that is most descriptive of the object. If possible both the horizontal and vertical location of a feature should be dimensioned in the same view. The upper part of Figure 5.4 shows an object properly dimensioned in millimeters to show the relation of the dimensions to the features in the most obvious manner. In the lower part of Figure 5.4, the dimensioning has several problems. The overall dimensions of the L-shaped profile are shown on the view where it is not clear that the object is L-shaped. The diameter of the hole is shown in a view where the hole is shown only as a pair of parallel hidden lines, not as a circle. And the vertical location of the hole is shown in one view, while the horizontal location is shown in another view.

The conventions for millimeter and inch dimensioning are different. For millimeter dimensioning, dimensions less than 1 mm have a zero preceding the decimal point, and integer dimensions may or may not include a decimal point and following zeros. Thus, legitimate millimeter dimensions are 0.8, 6, 8.00, and 1.5. Inch dimensions do not include a zero before the decimal point for values less than 1 inch. Zeros to the right of the decimal point are usually included. Legitimate inch dimensions are .800, 6.00, 8.00, and 1.500. For both millimeter and inch dimensions, the number of digits to the right of the decimal point may be used to indicate the tolerance (acceptable variation). It is standard practice to omit the units for dimensions. However, it is wise to note the units in the title block or as a note on the drawing.

correct

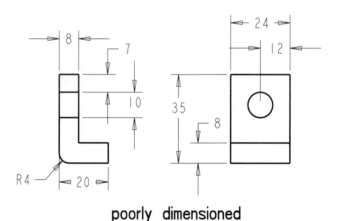

poorly dimensioned

Figure 5.4.

Circles and arcs are dimensioned using special symbols and rules. A radius is denoted using a leader line with an arrow pointing at the arc as shown in Figure 5.4. Sometimes a small cross is placed at the center of the radius. The position of the center should be dimensioned unless the radius is for a rounded edge, in which case the arc is positioned by virtue of being tangent at its ends to the sides forming the corner that is rounded, as shown in Figure 5.4. The dimension for a radius begins with a capital letter R to indicate radius. Full circles are dimensioned using the diameter with a leader line having an arrow pointing at the circle, as shown in Figure 5.4. If space permits, the leader line extends across a diameter of the circle with arrows at each end of the diameter. The dimension for a diameter begins with the Greek letter phi (Ø) to indicate diameter. The location of the center of the circle must also be dimensioned. Sometimes maintaining clarity makes it necessary to dimension several concentric circles from a side view, where the circle does not appear as a circle, thus violating the rule to dimension a feature in its most characteristic shape.

Angles are dimensioned as decimals or in degrees and minutes (60 minutes is one degree), so that 32.5° and 32°30′ are equivalent, where ′ indicates minutes. The dimension line for an angle is drawn as an arc with its center at the apex of the angle. When lines are drawn at right angles, a 90° angle is implied.

Multiple features that are identical need not be individually dimensioned. For example, the diameter of the two 3-mm holes in Figure 5.3 could be dimensioned with a leader line to one hole with the dimension 2× Ø3. The 2× indicates "two times." Alternative dimensions are Ø3 (2×), Ø3 2 PLACES, or Ø3 2 HOLES.

There are several other "rules" for dimensioning:

- Dimension lines should be outside of the outline of a part whenever possible.

- Dimension lines should not cross one another.

- Dimensions should be indicated on a view that shows the true length of the feature. This is particularly important when dimensioning features on a surface that is at an angle to the plane of the orthographic view.

- Each feature should be dimensioned only once. Do not duplicate the same dimension in different views.

- Dimension lines should be aligned and grouped when possible to promote clarity and uniform appearance. See for example the placement of the horizontal dimensions in Figure 5.3c.

- The numerical dimension and arrows should be placed between the extension lines where space permits. If there is space only for the dimension, but not the arrows, place the arrows outside of the extension lines. When the space is too small for either the arrows or the numerical value, place both outside of the extension lines. See Figure 5.4 for examples.

- Place the dimension no closer than about 10 mm (3/8 inch) from the object's outline.

- Dimensions should be placed in clear spaces, as close as possible to the feature they describe.

- When dimensions are nested (such as the horizontal dimensions in Figure 5.3c) the smallest dimensions should be closest to the object.

- Avoid crowding dimensions. Leave at least 6 mm (1/4 inch) between parallel dimension lines.

- Extension lines may cross visible lines of the object.

- Dimensions that apply to two adjacent views should be placed between the views, unless clarity is enhanced by placing them elsewhere. The dimension should be attached to only one view. Extension lines should not connect two views.

- Dimension lines and extension lines should not cross, if possible. Extension lines may cross other extension lines.

- A centerline may be extended to serve as an extension line, in which case it is still drawn as a centerline.

- Centerlines should not extend from view to view.

- Leader lines are usually sloped at about 30°, 45°, or 60° and are never horizontal or vertical.

- Numerical values for dimensions should be centered between arrowheads, except in a stack of dimensions (like Figure 5.3c).

- When a rough, noncritical dimension (such as a round) is indicated on a drawing, add the note TYP to the dimension to indicate that the dimension is "typical" or approximate.

5.5   SECTION AND DETAIL VIEWS

The cutaway view of a device appeared first in various forms in the 15th and 16th centuries to show details of parts hidden by other elements. These cutaway views have evolved to *section views* in which interior features that cannot be effectively displayed by hidden lines are exposed by slicing through a section of the object. To create a section view, a cutting plane is passed through the part and the portion of the part on one side of the cutting plane is imagined to be removed. In a section view, all visible edges and contours behind the cutting plane are shown. Hidden lines are usually omitted. The portion of the object that is sliced through is designated with angled crosshatch lines known as section-lining.

Key to the interpretation of a section view is the clear representation of where the cutting plane is and from which direction it is viewed. This is accomplished using a *cutting plane line* (long dash - short dash - short dash - long dash with perpendicular arrows at each end). The cutting plane line shows where the cutting plane passes through the object. The arrows on either end of the cutting plane line show the direction of the line of sight for the section view. An example of a block with a hole that has a large diameter partway into it and a smaller diameter through it is shown in Figure 5.5. In this case, the object is sectioned along a plane parallel to the right and left edges of the front view and through the center of the hole, as indicated by the position of cutting plane line A-A. The arrows indicate that the section view will be shown as if we are looking in the direction of the arrows. Thus, the section view shown to the right of the front view (as if unfolding the glass box) shows what the viewer would see if the portion of the object to the right of the cutting plane line was removed. The viewer sees the two diameters of the hole with the material that has been "cut" shown as cross-hatched. This view is designated Section A-A to correspond to the cutting plane line A-A. This type of section in which a single plane goes completely through an object is known as a *full section*.

Figure 5.5.

A slightly more complicated example of a full section is shown in Figure 5.6. The whole object is shown in the upper left portion of the figure, and the sectioned object is shown with a portion removed just below it. The right side of the figure shows a top view of the object with the cutting plane displayed. Imagine that the material on the side of the cutting plane in the direction the arrows are pointing is retained, and the material behind the arrows is removed. Then the projection of the retained portion appears as SECTION B-B when viewed in the direction of the arrows.

Figure 5.6.

SECTION B-B

Much more sophisticated section views can be created for complicated parts. In all cases, though, the clear definition of the cutting plane line and the direction of the line of sight for the view are critical. For example, Figure 5.7 shows a *half section* of the same part. The cutting plane extends halfway through the object to show the interior of one half of the object and the exterior of the other half. This type of section is ideal for

Section L-L

Removed Portion

SECTION   L-L

Figure 5.7.

symmetrical parts in which it is desired to show internal and external features in a single view. An *offset section* like that shown in Figure 5.8 is used to show internal details that are not in the same plane. The cutting plane for offset sections is bent at 90° angles to pass through important features. Note that the change of plane that occurs at the

Section H-H

Removed Portion

SECTION H-H

Figure 5.8.

90° bends is not represented with lines on the section view. When only a portion of the object needs to be sectioned to show a particular internal detail, a *broken-out section* like that shown in Figure 5.9 can be used. A break line separates the sectioned portion from the unsectioned portion, but no cutting plane line is drawn.

Figure 5.9.

In some cases it is necessary to clarify details of the part that may not be readily evident in the views normally shown in the drawing. For instance, a particular detail may be so small that it cannot be shown clearly on the same scale as the remainder of the part. This is done using an auxiliary *detail view* consisting of a small portion of the object magnified to make clear the small feature. An example is shown in Figure 5.10. In this case the small rivet holding the blade onto the arms of the pizza cutter is too small to see in the orthographic views. The region of interest near the rivet is circled in the left-most projection of the pizza cutter, and a note directs the reader to a magnified detail view in the upper right part of the drawing. The cross section of the rivet is clearly visible in this detail view.

Figure 5.10.

5.6 FASTENERS AND SCREW THREADS

Fasteners include a broad range of items such as bolts, nuts, screws, and rivets used to "fasten" parts together. In most cases, fasteners are standard parts purchased from an outside vendor, so detail drawings of fasteners are rarely necessary. Nevertheless, threaded holes and threaded shafts must be represented on a drawing.

The geometry of screw threads is too complicated to draw exactly in an engineering drawing and screw threads are standard, so either of two simple conventions is used to indicate screw threads, as shown in Figure 5.11. The *schematic representation* is used when a realistic representation of the side view of a screw thread is desired. For an external thread, the lines that extend across the entire diameter represent the *crest*, or peak, of the thread, while the shorter lines in between represent the *root*, or valley, of the thread. The distance between crests or between roots is called the *pitch*. On a drawing, the crests and roots are shown perpendicular to the axis of the threaded section rather than helical as they actually appear on the physical thread. In the end view, threads are depicted with concentric circles. The outer circle is the largest diameter of the screw thread, known as the *major diameter*, and the inner circle is the smallest diameter of the screw thread, or *minor diameter*. In the end view of the external thread, both diameters are shown as solid circles. In the *simplified representation* of external screw threads, the threads are omitted altogether. The major diameter is represented as a solid line in the side view, and the minor diameter is represented with a dashed line.

Figure 5.11.

The end view of an internal thread is shown as two concentric circles representing the major and minor diameter, but the major diameter is a hidden line. For a side view of a hidden internal screw thread, the major and minor diameters are both shown as hidden lines for both schematic and simplified screw thread representations. In a cross-section, internal threads can be shown in either the schematic or the simplified representation, as shown. For a hole that does not go completely through the part, called a *blind hole*, the lines in the side view representing the minor diameter continue deeper than the thread and come to a point. This represents the hole that is drilled prior to tapping, or cutting threads in the hole. The hole needs to be longer than the threaded section to permit the tool that is used to cut the threads, or tap, to penetrate deeply enough to fully cut the threads in the portion of the hole to be tapped.

Screw threads are specified in terms of the nominal (major) diameter, the pitch, and the thread series. For instance, the designations 1/4-20 UNC or .25-20 UNC or 1/4-20 NC all indicate a major diameter of .25 inches, a pitch of 20 threads per inch, and the United Course series. The diameter of the hole drilled before tapping an internal thread is specified in a wide variety of machinist and engineering handbooks. In this case the proper hole to be drilled has a diameter of .2010 inches, which corresponds to a number 7 drill. To make an internal thread of this size, a machinist would first drill a hole using a number 7 drill bit and then cut the threads using a 1/4-20 tap. For screw threads with a nominal diameter less than .25 inches, a number designation is used to specify the nominal diameter. For example, the designation 10-32 UNF indicates a nominal diameter of .1900 inches, 32 threads per inch, and the United Fine series. The pitch is equal to 1 divided by the number of threads per inch.

Metric threads are specified in a slightly different way. A designation M10 × 1.5 indicates a metric thread with a 10 mm major diameter and a pitch of 1.5 mm between crests of the thread. The number of threads per mm is 1 divided by the pitch. Other letters and numbers may follow the thread specification to denote tolerances and deviations of the thread, but for many cases these are not necessary. In addition, for a blind hole, the depth of the thread can be specified in several ways. Following the thread designation, the notation X .50 DEEP, THD .50 DP, or a downward arrow with a horizontal bar at its tail followed by .50 all indicate that the thread should extend .50 inches below the surface of the piece, as shown in Figure 5.11. (THD indicates "thread," and DP indicates "deep.") The threaded depth is usually 1.5 to 2 times the nominal diameter of the screw thread.

The dimensions for a threaded hole are indicated on a drawing using a leader line with the arrow pointing at the major diameter, as shown in Figure 5.11. Threaded holes are usually dimensioned in the view where they appear as circles, rather than in a side view of the threaded hole. Multiple threaded holes of the same specification are typically denoted in the same way as other multiple features using the notation (2X) or 2 TIMES at the end of the thread designation.

Nuts and bolts are usually only shown in a parts list and not on any drawings. They are specified in the same way as threaded holes, plus a bolt length and the type of head. For instance .25-28 UNF X 1.50 HEXAGON CAP SCREW or .25-28 NF X 1.50 HEX CAP SCR indicates a bolt that is 1.5 inches long with hexagonal head. Nuts are specified in terms of their thread and shape, such as M8 X 1.25 HEX NUT.

5.7   ASSEMBLY DRAWINGS

An assembly drawing shows all of the components of a design either assembled or in an exploded view. Many times assembly drawings include sections. Most dimensions are omitted in assembly drawings. Individual parts are not dimensioned, but some dimensions of the assembled mechanism may be included. Hidden lines are seldom necessary in assembly drawings, although they can be used where they clarify the design. Leader lines attached to a ballooned letter or detail number, as shown in Figure 5.12, reference the parts of the assembly. The leader lines should not cross and nearby leader lines should be approximately parallel. Sometimes parts are labeled by name rather than number. The parts list may be on the assembly drawing (usually on the right side or at the bottom) or it may be a separate sheet. The assembly drawing may also include machining or assembly information in the form of notes on the drawing.

	QTY.	PART NO.	DESCRIPTION
1	1	52806	handle
2	1	52825	cap
3	1	42886	guard
4	1	97512	rivet
5	2	55654	arm
6	1	56483	blade

Figure 5.12.

Often assembly drawings include assembly sections. These are typically orthographic or pictorial section views of parts as put together in an assembly. Adjacent parts in assembly drawings are cross-hatched at different angles to make the separate parts clear. The assembly cross-section on the left side of Figure 5.10 shows the interior structure of the parts of a pizza cutter and how they fit together. The detail view in the upper right of Figure 5.10 shows the cross-section through the rivet to indicate how the parts are assembled. Usually standard parts such as fasteners, washers, springs, bearings, and gears are not cross-hatched. But in the case of Figure 5.10, the rivet detail is integral to the assembly, and it is helpful to show its cross-section with cross-hatching to indicate how it fits with the other parts.

KEY TERMS

Centerlines	Dimension lines	Scale
Contour dimensioning	Extension lines	Section views
Cutting plane lines	Hidden lines	Title block
Detail view	Leader lines	Visible lines

Problems

1. Three orthographic views and an isometric view of an object that has overall dimensions of $4 \times 4 \times 4$ inches are to be shown in a drawing. Determine the best ANSI sheet size for an object of this size at the scales indicated below:

    a.   Half Size.

    b.   2:1.

    c.   1 = 1.

2. Three orthographic views and an isometric view of an object that has overall dimensions of $200 \times 200 \times 200$ mm are to be shown in a drawing. Determine the optimal standard metric scale to be used for the ISO sheet sizes indicated below:

    a.   A1.

    b.   A3.

    c.   A4.

3. The drawing shown in Figure 5.13 has many errors in the dimensioning. Sketch the orthographic views and show proper placement of dimensions.

Figure 5.13.

4. The drawing shown in Figure 5.14 has many errors in the dimensioning. Sketch the orthographic views and show proper placement of dimensions. The M10 × 1.5 threaded hole requires a tap drill of 8.5 mm. Only two views are needed for this part.

Figure 5.14.

5. Sketch the following section views for the object shown in Figure 5.15. The threaded holes extend from the top surface approximately halfway through the thickness of the part. The other holes extend through the part.

a. A-A.
b. B-B.

Isometric View   Top View

Figure 5.15.

6. Sketch the following section views for the object shown in Figure 5.16. All holes extend through the part.

    a. B-B.
    b. D-D.

Figure 5.16.

7. Look up the following machine screw threads in an engineering graphics book (such as *Engineering Graphics* by Giesecke, et al), a machinists or engineering handbook (such as *Marks' Standard Handbook for Mechanical Engineers*), an industrial supply catalog (such as McMaster–Carr), or on the World Wide Web (search on "tap drill size"). Specify the series (UNC or UNF), tap drill size or number, number of threads per inch, pitch, and nominal diameter.

    a. 4-40.
    b. 6-32.
    c. 8-32.
    d. 3/8-24.
    e. 6-40.
    f. 1/4-28.

8. Look up the following machine screw threads in an engineering graphics book (such as *Engineering Graphics* by Giesecke, et al), a machinists or engineering handbook (such as *Marks' Standard Handbook for Mechanical Engineers*), an industrial supply catalog (such as McMaster–Carr), or on the World Wide Web (search on "tap drill size metric"). Specify the series (course or fine), tap drill size, number of threads per mm, and nominal size.

    a. M2 × 0.4.
    b. M12 × 1.75.
    c. M30 × 3.5.
    d. M30 × 2.
    e. M10 × 1.25.

6

Tolerances

OVERVIEW

Tolerances on dimensions are necessary to specify the acceptable variability in the dimension of a part. Tolerances can be indicated for all dimensions using a general note or they can be specified for each individual dimension. Tolerances are based on the function of the part or are used to assure that mating parts fit together. However, in many cases the manufacturing process determines the tolerance. The surface finish may also be specified on a drawing to indicate the roughness of the surface. To further minimize ambiguity in engineering drawings, geometric dimensioning and tolerancing is used to specify the form of a feature, such as flatness or roundness, or to indicate the ideal position of a feature.

6.1 WHY TOLERANCES?

In 1815, it was hoped that muskets could be produced in a number of different U. S. Government and private armories so that the parts of the musket would be interchangeable. Before this time, muskets were individually handcrafted, so parts from one musket rarely fit another musket. Because engineering drawings were not commonly used, a number of "perfect" muskets were made according to master gages and jigs. *Gages* are devices used to check the individual

SECTIONS

- 6.1 Why Tolerances?
- 6.2 Displaying Tolerances on Drawings
- 6.3 How to Determine Tolerances
- 6.4 Surface Finish
- 6.5 Geometric Dimensioning and Tolerancing

OBJECTIVES

After reading this chapter, you should be able to:

- Explain interchangeable parts
- Explain the necessity of tolerancing
- Read a general tolerance on a drawing
- Read limit dimensions and tolerances on individual dimensions
- Determine a tolerance stackup
- Determine the range of tolerances for manufacturing processes
- Read a surface finish specification
- Understand basic form tolerances
- Understand simple positional tolerances
- Read geometric dimensioning and tolerancing symbols

dimensions of a part. For instance, the diameter of a shaft could be checked using a gage consisting of two holes: one of the maximum allowable diameter of the shaft and one of the minimum allowable diameter. If the shaft fit through the larger hole but not the smaller hole, then it was within the allowable dimensions for the shaft. *Jigs* are devices used to hold a workpiece and guide the tool to assure repeatable machining. The master gages and jigs were sent to the armories with the instruction that "no deviations from the pattern were to be allowed." In 1824, this concept of *interchangeable parts* was tested by disassembling a hundred rifles from several different armories, mixing the parts, and then reassembling at random. Most reassembled rifles worked as they should, proving the concept and value of interchangeable parts.

The interchangeability of parts is so common now that it is hard to imagine anything different. The concept allows parts made in various locations by different manufacturers to be successfully assembled and to function properly as an assembly. Although engineering drawings rather than "master parts" came into use in the mid-19th century, it was the two world wars in the 20th century that brought about the development of methods of showing the acceptable variation in a dimension on a drawing, or *tolerancing*

The need to control precisely the geometry of a part arises from the part's function. For instance, the cross-sectional geometry of an airfoil, such as the wing of a jet aircraft or a turbine blade, must be accurately controlled to assure aerodynamic efficiency. The more commonplace need for controlling geometry results from the requirement for parts to fit together. But it is quite difficult to make every part the exact size that is specified in a dimension because of slight differences in tool size, machine tool wear, human operator error, and other factors. As a consequence, tolerances are used in engineering drawings to specify the limit in the variation between mating parts and to provide guidelines to control the manufacturing process. The *tolerance* is the amount that a specific dimension is permitted to vary, or the difference between the maximum and minimum limits of a dimension.

Consider a dimension on a drawing specified as 3.750 ± .003. This dimension indicates that the part has a *nominal size*, or general size, of 3 3/4 inches (usually expressed as a common fraction). The *basic size*, or theoretical size for the application of the tolerance, is 3.750 inches (expressed as a decimal). The tolerance of the dimension is .006 inches. Thus, an acceptable machined part can have an *actual size* ranging from 3.747 to 3.753 inches, which are the minimum and maximum limits of the dimension. If the actual size is smaller than 3.747 or larger than 3.753, then the part is not acceptable.

Of course, it is desirable to make the part as close as possible to the basic size, sometimes called the *target size*. But the more accuracy needed in a machined part, the higher the manufacturing cost. Furthermore, the method of machining a part limits the tolerance that can be specified. For instance, the tolerance in drilling a .500 inch hole is between .002 and .005 depending on the quality of the drill press and drill bit. To require a tolerance smaller than .002 would require an additional machining process, such as reaming. Consequently, tolerances can play a large role in the cost of manufacturing a part. Therefore, the tolerances should not be specified tighter than necessary for the product to function properly.

An increased awareness of the importance of tolerances in manufacturing has led to an approach in which the deviation of the actual size of a part from the target size has a cost in terms of reworking of the part, scrap, customer dissatisfaction, and poor reliability. An approach to handle this is to design a product so that it is less sensitive to manufacturing tolerances. The idea is to design the part intelligently so that easy-to-manufacture tolerances, rather than exceptionally tight tolerances, are necessary. Consequently, proper fit between mating parts comes about because of good design, instead of depending on tight tolerances. This approach is known as *robust design*.

6.2   DISPLAYING TOLERANCES ON DRAWINGS

Every dimension on a drawing should have a tolerance. Tolerances can be displayed on engineering drawings in several ways depending on the situation.

General tolerances can be given in a note on the drawing or in the title block. An example would be the note ALL DIMENIONS TO BE HELD TO ± .003″. Thus, a dimension of .375 would have a minimum limit of .372 and a maximum limit of .378. Often general tolerances are specified in terms of the number of digits following the decimal point in a dimension. In this case, a note would appear on the drawing such as:

$$\begin{aligned}
&\text{UNLESS OTHERWISE SPECIFIED}\\
&\text{TOL:.XX} = \pm.010\\
&\text{.XXX} = \pm.005\\
&\text{HOLES} = \pm.002\\
&\text{ANGLES} = \pm.5\ \text{DEG}
\end{aligned} \tag{6-1}$$

Thus, a dimension of .50 on a drawing would indicate an actual size between .490 and .510. A dimension of .500 would indicate an actual size between .495 and .505. For metric dimension, a similar form for general tolerances would be X.X METRIC = ±.08.

Two other methods of displaying tolerances on a drawing are shown in Figure 6.1. The maximum and minimum sizes are specified directly when using *limit dimensions* to specify the tolerance. Usually the upper limit is placed above the lower limit. While this method clearly shows the limits of the dimension, many engineers, designers, and machinists think in terms of the nominal or basic size plus a tolerance, neither of which are immediately evident using limit dimensions. Thus, a more common approach is to use *plus and minus dimensions* where the basic size is shown followed by the tolerance values. If the plus and minus tolerance values are identical, then the basic dimension is followed by a plus/minus sign and half of the numerical value of the tolerance. Otherwise, the plus and minus tolerances are indicated separately with the plus above the minus. A *bilateral tolerance*, such as the tolerance for the 1.500 inch dimension in Figure 6.1, varies in both directions from the basic size. A *unilateral tolerance*, such as the tolerance on the shaft in Figure 6.1, varies in only one direction from the basic size. In both cases, the number of digits to the right of the decimal point for the dimension should be the same as the number of digits to the right of the decimal point for the tolerance.

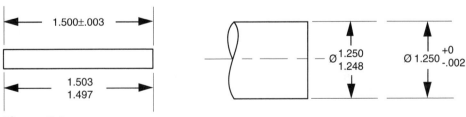

Figure 6.1.

Sometimes it is not necessary to specify both limits of a tolerance and only one limit dimension is needed. MIN or MAX is placed after a numerical dimension to indicate that it is a *single limit dimension*. The depth of holes, length of threads, and radii of corner rounds are often specified using single limit dimensions.

Many times several methods of representing tolerances are used on a single drawing. For instance, general tolerances may be listed in the title block and plus and minus dimensions or limit dimensions may be used for a few dimensions to which the general tolerances do not apply.

An important consideration in applying tolerances is the effect of one tolerance on another, especially because tolerances are cumulative. This is known as *tolerance accumulation* or *tolerance stackup* for a chain of dimensions. In the example shown in Figure 6.2, two dimensioning schemes are shown. Above the part, *chain dimensioning* shows the distance from one feature to the next. This would be an appropriate dimensioning scheme if the distance from one hole to the next were critical, because the distance between holes must be between .995 and 1.005. But if the distance of the holes from left edge of the part were critical, a large tolerance would accumulate. The third hole would be 2.985 from the left edge if all of the dimensions happened to be at the minimum limit of the dimension, or 3.015 from the left edge if all dimensions were at the maximum limit. Thus, the actual dimension of the third hole from the left edge using chain dimensioning is 3.000 ± .015. This tolerance accumulation could be avoided using *datum dimensioning*, where the dimensions are given with respect to a datum, in this case the left edge of the part as shown below the part in Figure 6.2. Using this scheme the third hole is specified to be between 2.995 and 3.005 from the left edge. Of course, the distance between the second and third holes could be as small as .990, if the second hole is at 2.005 and the third hole is at 2.995. The distance between the second and third holes could be as large as 1.010 if the holes are at 1.995 and 3.005. Consequently, the decision about which way to dimension the holes depends on whether the distance between holes is critical, or whether the distance from each hole to the left edge is critical.

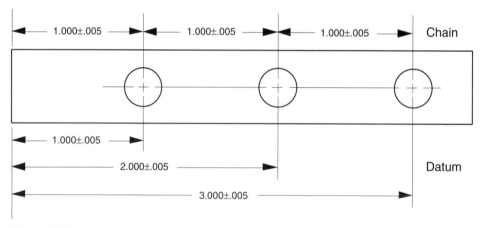

Figure 6.2.

6.3   HOW TO DETERMINE TOLERANCES

One of the most difficult aspects of dimensioning and tolerancing for inexperienced engineers and designers is to determine what tolerance to specify on a drawing. In some cases a specific tolerance is quite clear based on the function of the design. For instance, if a 3/4-inch shaft must rotate at moderate speed in a hole, the tolerances on the shaft and the hole can be determined based on standard classes of fits that are available in many engineering and machinist handbooks. In this particular case the hole could be between .7500 and .7512, while the shaft could have a diameter of .7484 to .7492. A shaft any bigger or a hole any smaller would result in too tight of a clearance for the shaft to rotate freely, or could result in the shaft being too big for the hole. A smaller shaft or larger hole than the specified limits would result in a sloppy fit between the shaft and the hole. The point here is not the details of how the tolerances are determined. That is too specific to individual cases for this discussion and is presented in many handbooks and texts. Instead, the point is that the function of the system that is being designed drives the tolerances. In this case, *the tolerances drive the manufacturing process* used to machine the hole and the shaft.

But more a more common situation is the opposite: *the manufacturing process drives the tolerance*. If the only machine tool that is available to form a hole is a drill press, the 3/4-inch hole that will result will be between .748 and .754 in diameter based on expected tolerances for the drilling operation. But this means that the hole in the example above would have too much variability to assure that the shaft rotates in it properly. Of course, if only one shaft-hole assembly was being made, the hole could be drilled first and measured. Then the shaft could be turned on a lathe to a diameter just slightly smaller to achieve the fit described above. But this approach will clearly not work for mass-produced, interchangeable parts. Either a more accurate machining method is needed to create the hole or the design must be altered to avoid drilling a tightly dimensioned hole altogether. (In this example, a standard-size, off-the-shelf bearing could be inserted into a larger drilled hole to mate with the shaft.)

When the manufacturing processes that are available drive the tolerances, the tolerances on the drawing must reflect the manufacturing process that is used to make the part, and the design must be altered to ensure that the part can be made with the available machining processes. Guidelines are available for the tolerance range of various common machining processes. These processes include the following:

Lapping: A process to produce a very smooth and accurate surface by rubbing the surface of the workpiece against a mating form, often using a fine abrasive between the surfaces.

Honing: A process in which a honing stone made of fine abrasive is used to form a surface on a workpiece.

Grinding: A process to remove a small amount of material from a workpiece with a rapidly rotating grinding wheel.

Broaching: A process similar to filing in which a cutting tool is reciprocated along its axis to remove a small amount of material from a workpiece.

Reaming: Removal of a small amount of material from a drilled hole using a rotating cutting tool (reamer).

Turning: Axisymmetric removal of material from a rotating workpiece by feeding a cutting tool into the workpiece using a lathe.

Boring: Axisymmetric removal of material from a workpiece using a tool with a single cutting surface. Either the workpiece or the tool may be rotating.

Milling: Removal of material by feeding workpiece into a rotating cutting tool.

Drilling: Creating a hole by feeding the end of a sharpened cylindrical cutting tool (drill bit) into the material.

Stamping and *Punching*: Cutting material using a punch and a die to shear the material, much like a paper hole-punch.

The tolerance typically depends on the size of the feature, as well as the quality of the machine tool, sharpness of the cutting tool, and skill of the operator. Figure 6.3 indicates the range of tolerances to be expected for various machining operations. To use this table, first consider the size of the feature in the left column. Then go below the sizes in the left column to the type of machining operation used. The bar associated with that process indicates the range of tolerances in the row associated with the feature size. For instance, consider a feature that has a dimension of 2 inches. If broaching were used to create that feature, the chart indicates that the total tolerance would be .0004 to .0015, depending on the machining conditions. Thus, this dimension could be specified as 2.0000 ± .0002 up to 2.0000 ± .0008 (rounding up). If the same piece was created on a milling machine, the tolerances would be .0025 to .010, indicating a dimension of 2.0000 ± .0013 up to 2.000 ± .005. If broaching were not available, we would have to be satisfied with the tolerances of a milling machine and adjust the design accordingly, if the milling tolerances are too large. If both broaching and milling were available, the criticality of the tolerance and the cost of manufacturing would drive the tolerance that would be specified on the drawing. (Broaching, a mass-production process, is usually more expensive than milling, so milling would be preferred if the milling tolerances are acceptable.) Tolerances for metric dimensioning can be determined by converting the values in Figure 6.3 to millimeters and rounding off the result to one less place to the right of the decimal point.

Size (in.)	Total Tolerance (in.)								
0.000-0.599	0.00015	0.0002	0.0003	0.0005	0.0008	0.0012	0.002	0.003	0.005
0.600-0.999	0.00015	0.00025	0.0004	0.0006	0.001	0.0015	0.0025	0.004	0.006
1.000-1.499	0.0002	0.0003	0.0005	0.0008	0.0012	0.002	0.003	0.005	0.008
1.500-2.799	0.00025	0.0004	0.0006	0.001	0.0015	0.0025	0.004	0.006	0.010
2.800-4.499	0.0003	0.0005	0.0008	0.0012	0.002	0.003	0.005	0.008	0.012
4.500-7.799	0.0004	0.0006	0.001	0.0015	0.0025	0.004	0.006	0.010	0.015
7.800-13.599	0.0005	0.0008	0.0012	0.002	0.003	0.005	0.008	0.012	0.025

Operation

lapping/honing
grinding/burnishing
broaching
reaming
turning/boring
milling
stamping/punching

Figure 6.3.

Typical tolerances on drills depend on the drill size. Standard drill sizes are fractional in increments of 1/64-inch from 1/16-inch to 4-inch and according to an alphanumeric code. The coded drill sizes range from 80 (.0135 inch) to 1 (.2280 inch) and from A (.234 inch) to Z (.413 inch). The dimensions for the coded drill sizes are widely available in engineering and machinist handbooks. In addition, the proper drill sizes for drilling holes that will be tapped are listed in these handbooks. The standard tolerance for drilled holes is given in Table 6.1.

TABLE 6-1 Standard Tolerance For Drilled Holes

DRILL SIZE (IN.)	TOLERANCE (IN.)	
	Plus	Minus
0.0135 (80)–0.185 (13)	0.003	0.002
0.1875–0.246 (D)	0.004	0.002
0.250–0.750	0.005	0.002
0.7656–1.000	0.007	0.003
1.0156–2.000	0.010	0.004
2.0312–3.500	0.015	0.005

6.4   SURFACE FINISH

All surfaces are rough and irregular on a microscopic scale. The roughness can be described in terms of asperities (peaks or ridges) and valleys. The irregularities of a surface are classified into *roughness* and *waviness*. Roughness describes the small scale, somewhat random irregularity of a surface, while waviness describes a more regular variation in the surface that typically has a wavelength of 0.8 mm or more. The surface can be thought of as a roughness superimposed on waviness. The roughness and waviness can be measured in several ways, but the most common is a profilometer, which drags a fine-pointed stylus over the surface and records the height of the stylus. The roughness is measured in terms of the roughness average (R_a). The roughness average is defined as the average of the absolute values of the deviation of the local surface height from the average height of the surface. This is equivalent to half the average of the peak to valley height. In addition, the *lay*, or predominant direction of the marks left by the machine tools, may be important to describing the surface.

The surface finish resulting from several machining and forming processes are shown in Figure 6.4. This figure shows the range of possible surface roughness averages (R_a) in both micrometers (1μm = 10^{-6} m) and microinches (1μin = 10^{-6} in). For instance, surfaces formed by drilling are likely to have a roughness from 12.5 down to 0.8 μm (500 down to 32 μin). Typically, the roughness is in the middle of the specified range, about 6.3 down to 1.6 μm for the case of drilling. In addition to the machining processes described in Section 6.3, the following processes are also included in Figure 6.4:

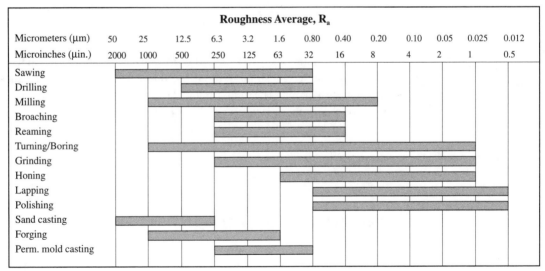

Figure 6.4.

Sawing: Cutting material by moving a toothed blade through the material.

Polishing: Removal of scratches and tool marks on a surface using a belt or rotating wheel of soft material such as cloth or leather, often with a fine abrasive.

Sand Casting: Forming a part by pouring molten metal into a mold made of sand.

Forging: Shaping of metal to a desired form by pressing or hammering, usually with the metal hot.

Permanent Mold Casting: Forming a part by pouring molten metal into a permanent mold.

To put the roughness averages in perspective, consider that a 0.10 µm roughness is a mirrorlike surface, free from visible marks of any kind. A 1.6 µm roughness is a high-quality, smooth machined finish. A 6.3 µm roughness is the ordinary finish from typical machining operations.

The surface finish may be specified on a drawing using the special symbols touching the surface or with a leader line to the surface as shown in Figure 6.5. For the top surface, the roughness is specified as 1.2 µm, corresponding to a high quality, smooth machine finish. The surface finish symbol on the right side of the object includes a horizontal bar that indicates that material removal by machining is required for this surface. The surface finish symbol on the left side of the object indicates that material removal from this side is prohibited. This surface must be that resulting from the original process used to produce the surface such as casting, forging, or injection molding.

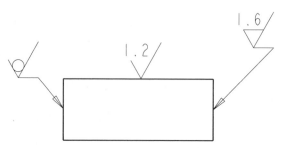

Figure 6.5.

Other notations can be added to the surface roughness symbol. A number to the left of the surface finish mark describes the material removal allowance indicating the amount of stock material to be removed by machining. A symbol to the right of the surface roughness mark indicates the lay, or predominant direction of machining marks: perpendicular, parallel, circular, or radial. Other notations can be used to specify the waviness. Details of surface finish notation are available in engineering and machinist handbooks.

6.5   GEOMETRIC DIMENSIONING AND TOLERANCING

Ideally, engineering drawings should be unambiguous in their interpretation. In many cases, stating dimensions and tolerances provides adequate information so that the part can be manufactured to ensure interchangeability of parts and optimal performance. But sometimes, the traditional dimensioning using the plus and minus scheme does not adequately describe the geometry of the part. For instance, consider the part shown in Figure 6.6. The dimension shown in the drawing could have several possible interpretations. For instance, the dimension at the top edge could be 14.9 mm while the dimension at the bottom edge could be 15.1 mm, resulting in a trapezoidal part instead of a rectangular part. Or the right edge of the part could be bowed so that the part is 15.1 mm wide at the top and bottom, but 14.9 mm wide at the middle. As a result, the part would not fit flush against another flat surface even though the part is 15 ± .1 mm wide at any particular location. In both of these cases, the geometry of the part is not described fully by the plus and minus tolerances on the dimension.

Figure 6.6.

A system known as *geometric dimensioning and tolerancing* is used to prevent such ambiguities in the design to present the design intent more clearly. Geometric dimensioning and tolerancing supplements the traditional dimensioning system with specifications that describe the geometry of the part to enhance the interpretation of the drawing. It is used to indicate dimensions critical to a part's function and ensure part interchangeability. Items related to geometric dimensioning and tolerancing are indicated on a drawing by a feature symbol consisting of a frame or box around information describing the geometry. A complete discussion of geometric dimensioning and tolerancing is beyond the scope of this book—there are entire books on the subject. Nevertheless, several of the key aspects and applications of geometric dimensioning and tolerancing are presented here to familiarize the reader with the basic terminology and symbols that are used.

The various types of tolerances and the characteristics that they describe are indicated in Figure 6.7. Each has its own symbol that is used on a drawing. Perhaps the easiest

	TYPE OF TOLERANCE	CHARACTERISTIC	SYMBOL
FOR INDIVIDUAL FEATURES	FORM	STRAIGHTNESS	—
		FLATNESS	▱
		CIRCULARITY (ROUNDNESS)	○
		CYLINDRICITY	⌯
FOR INDIVIDUAL OR RELATED FEATURES	PROFILE	PROFILE OF A LINE	⌒
		PROFILE OF A SURFACE	⌓
FOR RELATED FEATURES	ORIENTATION	ANGULARITY	∠
		PERPENDICULARITY	⊥
		PARALLELISM	//
	LOCATION	POSITION	⊕
		CONCENTRICITY	◎
		SYMMETRY	⚌
	RUNOUT	CIRCULAR RUNOUT	↗
		TOTAL RUNOUT	↗↗

Figure 6.7.

type of geometric tolerances to understand are the *form tolerances*, which describe the shape of a single feature. For instance, straightness indicates the limits on how much a surface or axis can bow with respect to a straight line. Figure 6.8 shows the *straightness* specification for a surface. A feature control box with the straightness symbol (a horizontal bar)

Drawing Notation

Interpretation

Figure 6.8.

and a numerical dimension are connected by a leader line to the surface. The interpretation is that the upper surface must be straight enough so that all points on the surface are within the tolerance zone of 0.010 mm. An example of a *flatness* specification is shown in Figure 6.9. The parallelogram symbol with a numerical dimension in a feature control box

Drawing Notation

Interpretation

Figure 6.9.

indicates that the planar surface must lie between two parallel planes 0.20 mm apart. Other single feature specifications include *circularity* or *roundness* (for circular features to be within a tolerance zone defined by two concentric circles), and *cylindricity* (for cylindrical features to be within a tolerance zone defined by two concentric cylinders). Closely related to form tolerances are *profile tolerances*, which define the acceptable deviation of the outline of an object (profile) from that specified in the drawing. These tolerances define a zone on either side of the true profile shown on the drawing within which the line or surface should remain. Profile tolerances can be applied to either a line or a surface.

Orientation tolerances prescribe the relation between features. *Parallelism* controls the degree to which a surface is parallel to a reference plane, or datum plane. In Figure 6.10, the horizontal datum plane that coincides with the bottom edge of the part is identified with a box around B attached to the bottom edge. The parallelism fea-

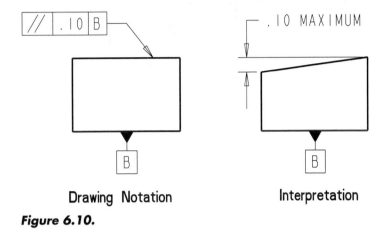

Drawing Notation   **Interpretation**

Figure 6.10.

ture symbol (parallel lines), the numerical tolerance, and the reference datum plane are included in a feature control box with a leader line to the surface to which the tolerance applies. In this case, the upper surface must be within .10 mm of parallel to datum plane B. *Perpendicularity* with respect to a datum plane is indicated in a similar way, as shown in Figure 6.11. Here the vertical plane must be perpendicular to datum plane A to

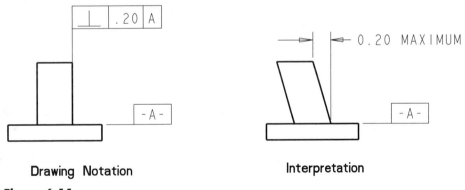

Drawing Notation   **Interpretation**

Figure 6.11.

within .20 mm. In this example, an alternative notation for specifying datum plane A is used. The datum plane symbol used in Figure 6.11 is common, but the datum plane symbol used in Figure 6.10 and Figure 6.12 is preferred. The datum plane symbol can be applied in several different ways as shown in Figure 6.10 and Figure 6.12. Several other relations between features can be specified, including the tolerance zone for *angularity* from a datum plane or axis, the tolerance zone for *concentricity* of a surface of revolution with respect to an axis, the tolerance zone for *symmetry* with respect to a datum plane, and the tolerance zone for *runout*, which describes the circularity with respect to an axis of revolution.

Figure 6.12.

The most difficult concept in geometric dimensioning and tolerancing is the concept of *positional tolerances*. It is demonstrated most easily in indicating the position of a hole, as shown in Figure 6.12a. In this case, the ultimate goal is to have a circular open area (hole) that is at least 25.0 mm in diameter with its center exactly at the position indicated in the drawing. Note that even though we want a 25.0 mm hole, the hole is dimensioned as 25.3 mm in the drawing. The reason for this will shortly become evident.

Rather than describing the position of the hole with a plus-or-minus tolerance, the ideal position of the hole is prescribed. The boxed dimensions indicate that they are basic *true-position dimensions*, which describe the perfect position of the center of the hole. Then the feature control box below the dimension of the hole shows how close the center of the hole should be to the true position. In this case, the hole should be within a circular tolerance zone with a diameter of .1 mm of the true position (indicated by the plus in the circle within the feature control box followed by ∅ .1). The reference datum planes are also included in the feature control box. Datum plane A is the surface of the part in the plane of the page. Datum planes B and C are references for the true-position dimensions of 40 and 60 mm. These three mutually perpendicular datum planes provide a reference point for the positional tolerance of the hole.

The positional tolerance zone for the center of the hole is shown in Figure 6.12b. The "true position of hole center" is based on the 40 and 60 mm dimensions. Around this center is a circle with a diameter of .1 mm (not drawn to scale in Figure 6.12b) that defines the "positional tolerance zone." The center of the actual hole must be within this circle. For instance, an acceptable hole would be a 25.1 mm diameter hole with its center on the edge of this circular tolerance zone, as shown in the Figure 6.12b.

The M in a circle in the feature control box in Figure 6.12a indicates that the .1 mm circular positional tolerance zone is applied at the *maximum material condition* (MMC). For a hole, the maximum material condition is the smallest acceptable dimension of the hole, because this is the situation in which the most material would be present. For the part in Figure 6.12a, the diameter of the hole for the maximum material condition is 25.1 mm, given a dimension of 25.3 ± .2 for the diameter of the hole. Thus, the circle representing a hole with its center on the edge of the positional tolerance zone has a diameter of 25.1 mm, as shown in Figure 6.12b. Of course, the 25.1 mm diameter hole could be anywhere within the positional tolerance zone or on its edge. Three more 25.1 mm diameter circles are drawn in Figure 6.12c in addition to the original 25.1 mm circle. The centers of all four circles are on the edge of the positional tolerance zone. Note that there is a circle that lies inside all of the 25.1 mm diameter circles, which we call the "theoretical hole." It is relatively straightforward to find the diameter of the theoretical hole, based on the positional tolerance zone for the center of the hole and the MMC size of the hole. The radius of the theoretical hole is the distance from the true position of the hole center to the nearest portion of any one of the MMC circles. Here the theoretical hole has a diameter of 25.0 mm. In fact the theoretical hole always has a diameter equal to the diameter for the maximum material condition minus the diameter of the positional tolerance (in this case, 25.1 mm − .1 mm = 25.0 mm).

The key result is that if the hole diameter is within the specified tolerance (25.3 ± .2 mm) and the center of the hole is within the positional tolerance zone (.1 mm), there will always be a circular opening in the part with a diameter of at least 25.0 mm. Now assume that the hole must mate with a circular pin that is positioned at the true center of the hole and is exactly 25.0 mm in diameter. Given the dimensioning scheme in Figure 6.12a, the pin will always align with the theoretical hole and fit through the theoretical hole, even if the actual hole is not positioned so its center is exactly at the true position of the hole center. Of course, the theoretical hole is based on the maximum material condition (smallest hole allowed). If the hole is at the large end of the acceptable tolerance, 25.5 mm in diameter, circles drawn in Figure 6.12c will be larger. But the theoretical hole will still be open for the pin. Likewise, if the circles drawn in Figure 6.12c are not on the edge of the positional tolerance zone, but inside of the zone, the entire theoretical hole will still be open for the pin. Thus, defining the position and diameter of the hole as in Figure 6.12a forces the existence of an open area (theoretical hole) that is at least 25.0 mm in diameter and centered at the true position of the hole center.

The maximum material condition reflects the practical aspect of assembly. Consider two parts that are bolted together with several bolts. If the bolt holes are not accurately positioned, not all of the bolts can be inserted. Either drilling out the holes to make them larger or using smaller diameter bolts fixes the problem. The idea of the maximum material condition is to capture these trade-offs. The maximum material condition indicates the inner boundary (the theoretical hole) for the actual hole. The boundary of the actual hole must always be outside of this inner boundary. In this way, a bolt or pin that is positioned at the true position of the hole will always fit through the actual hole. Likewise, the least material condition (LMC) , or largest hole size, can be used to define a theoretical circle within which the actual hole will be. The boundary for the actual hole must lie inside of the theoretical circle defined by the LMC. Further information regarding geometric dimensioning and tolerancing is provided in handbooks and graphics textbooks.

KEY TERMS

Actual size
Basic size
Form tolerances
General tolerances
Geometric dimensioning
  and tolerancing

Limit dimensions
Maximum material condition
Nominal size
Orientation tolerances
Plus and minus dimensions

Positional tolerances
Roughness
Target size
Tolerance stackup
True-position dimensions

Problems

1. The part shown in Figure 6.13 is made using a very accurate milling machine by a skilled machinist. The hole is reamed. Specify plus and minus tolerances for each dimension assuming that it is critical to hold all dimensions as close as possible to those specified. Use four digits past the decimal point to specify the tolerances, even though typical practice would be to specify only three digits past the decimal point. Dimensions are in inches.

Figure 6.13.

2. The part shown in Figure 6.13 is made by an unskilled student using an old milling machine in a university student shop. The hole is drilled. Specify plus and minus tolerances for each dimension that are reasonable for these conditions. Use four digits past the decimal point to specify the tolerances, even though typical practice would be to specify only three digits past the decimal point. Assume dimensions are in inches.

3. Sketch a view of the object shown in Figure 6.14. Specify the horizontal dimension of the steps using chain dimensioning from the left end. The length of each horizontal section from left to right is .50, .75, 1.25, and .25 inches. Base the plus and minus dimensions on the worst expected tolerances for milling. What are the maximum and minimum dimensions for the overall horizontal length of the object? Use four digits past the decimal point to specify the tolerances, even though typical practice would be to specify only three digits past the decimal point.

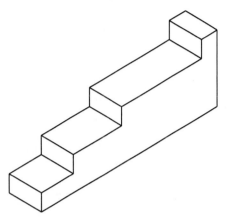

Figure 6.14.

4. Sketch a view of the object shown in Figure 6.14. Specify the horizontal dimension of the steps using datum dimensioning from the left end. The length of each horizontal section from left to right is .50, .75, 1.25, and .25 inches. Base the plus and minus dimensions on the best expected tolerances for milling. What are the maximum and minimum dimensions for the overall horizontal length of the object? Use four digits past the decimal point to specify the tolerances, even though typical practice would be to specify only three digits past the decimal point.

5. Sketch a side view of the shaft shown in Figure 6.15. Specify the horizontal dimension of each section of the shaft using chain dimensioning from the left end. The length of each horizontal section from left to right is .75, 2.00, .50, 1.25, and 3.00 inches. Base the plus and minus dimensions on the best expected tolerances for turning. What are the maximum and minimum dimensions for the overall horizontal length of the object? Use four digits past the decimal point to specify the tolerances, even though typical practice would be to specify only three digits past the decimal point.

Figure 6.15.

6. Sketch a side view of the shaft shown in Figure 6.15. Specify the horizontal dimension of each section of the shaft using datum dimensioning from the left end. The length of each horizontal section from left to right is .75, 2.00, .50, 1.25, and 3.00 inches. Base the plus and minus dimensions on the worst expected tolerances for turning. What are the maximum and minimum dimensions for the length from the left end to the left edge of the largest diameter section? Use four digits past the decimal point to specify the tolerances, even though typical practice would be to specify only three digits past the decimal point.

7. Draw the appropriate surface finish symbol for:

   a. surface roughness typical for drilling.
   b. surface roughness typical for grinding.
   c. the best possible surface roughness for milling with material removal required.
   d. surface finish of 12.5 microns with material removal prohibited.

8. Sketch the appropriate Geometric Dimensioning and Tolerance form tolerance for:

   a. an edge straight so that all points on the surface are within a tolerance zone of 0.020 mm.
   b. a surface that is flat to within 0.015 mm.
   c. a surface that is parallel to datum plane C to within 0.20 mm.
   d. a surface that is perpendicular to datum plane B to within 0.10 mm.
   e. a hole that is perpendicular to datum plane B to within 0.12 mm.

9. The part shown in Figure 6.16 is dimensioned in millimeters using the traditional coordinate method. Sketch the same views of the part and dimension using Geometric Dimensioning and Tolerancing procedures so the following functional requirements are met:

Figure 6.16.

- Set surfaces A and B as datum planes.
- Surface A is flat to within 0.15 mm.
- The upper edge in the front view from which the holes are located vertically is perpendicular to surface B to within 0.20 mm. Set this edge as datum plane C.
- The edge-forming surface A in the top view is straight to within 0.05 mm.
- The 10, 15, and 40 mm dimensions are true-position dimensions.
- The 5, 30, and 50 mm dimensions must be held to the tightest possible tolerance that can be achieved using milling. (Convert the dimensions to inches to determine the tolerances to two places after the decimal (round as necessary). Specify the tolerances as plus and minus dimensions in millimeters.)
- The holes have a tolerance of ±0.05mm. The theoretical hole size is 8.00mm.
- The holes are located within 0.15 mm of the true position at the Maximum Material Condition.

10. The part shown in Figure 6.13 is dimensioned in inches using the traditional coordinate method. Sketch the same views of the part and dimension using Geometric Dimensioning and Tolerancing procedures so the following functional requirements are met:

- Set surfaces A and B as datum planes. Set the right edge as datum plane D.
- Surface A is flat to within .005 inches.
- The hole is perpendicular to surface A to within .010 inches. (Place this tolerance just below the dimension.)
- Surface C is parallel to surface B to within .012 inches.
- The .75 and 1.00 dimensions for the location of the hole are true-position dimensions.
- The 1.50 and 3.35 dimensions must be held to the tightest possible tolerance that can be achieved using milling (use three digits past the decimal point for the tolerances).
- The hole has a tolerance of ±.002. The theoretical hole size is 1.000.
- The hole is located within .004 inches of the true position at the Maximum Material Condition.

11. The part shown in Figure 6.17 is dimensioned in millimeters using the traditional coordinate method. Sketch the same views of the part and dimension using Geometric Dimensioning and Tolerancing procedures so the following functional requirements are met:

- Set surface A as a datum plane.
- Surface B is flat to within 0.10 mm.
- Surface B is parallel to surface A to within 0.05 mm.
- The 40 and 100 dimensions need only be held to the loosest possible tolerance that can be achieved using milling. (Convert the dimensions to inches to determine the tolerances. Specify the tolerances as plus and minus dimensions in millimeters.)

Figure 6.17.

References

ASME Y14.5M-1994, Dimensioning and Tolerancing. New York, NY: American Society of Mechanical Engineers, 1995.

G. R. Bertoline, *Introduction to Graphics Communications for Engineers.* New York, NY: McGraw-Hill, 1999.

P. J. Booker, *A History of Engineering Drawing.* London, England: Northgate Publishing Co. Ltd., 1979.

S. H. Chasen, Historical highlights of interactive computer graphics. *Mechanical Engineering,* November 1981, 32–41.

G. E. Dieter, *Engineering Design: A Materials and Processing Approach.* New York, NY: McGraw-Hill, 1991.

A. R. Eide, R. D. Jenison, L. H. Mashaw, L. L. Northup, and C. G. Sanders, *Engineering Graphics Fundamentals.* New York, NY: McGraw-Hill, 1985.

J. Encarnacao and E. G. Schlechtendahl, *Computer Aided Design.* Berlin, Germany: Springer-Verlag, 1983.

E. S. Ferguson, *Engineering and the Mind's Eye.* Cambridge MA: The MIT Press, 1992.

E. S. Ferguson, The mind's eye: Nonverbal thought in technology. *Science* 197:827–836, 1977.

F. E. Giesecke, A. Mitchell, H. C. Spencer, I. L. Hill, R. O. Loving, J. T. Dygdon, J. E. Novak, and S. Lockhart, *Engineering Graphics.* Upper Saddle River, NJ: Prentice Hall, 1998.

K. Hanks, *Rapid Viz: A New Method for the Rapid Visualization of Ideas,* Menlo Park, CA: Crisp Publications, 1990.

C. M. Hoffman, *Geometric and Solid Modeling,* San Mateo, CA: Morgan Kaufmann Publishers, Inc., 1989.

S. Pugh, *Total Design,* Reading, MA: Addison Wesley, 1990.

D. G. Ullman, *The Mechanical Design Process,* 2nd Edition. New York, NY: McGraw-Hill, 1997.

D. G. Ullman, S. Wood, and D. Craig, The importance of drawing in the mechanical design process. *Computers and graphics,* 14:263–274, 1990.

W. R. D. Wilson, Course Notes for ME B40: Introduction to Mechanical Design and Manufacturing. Northwestern University, March 1998.

Index

A

a tolerance 69
Actual size 68
Albrecht Dürer 4
American National Standards Institute 50
Angles 54
angularity 77
Assembly drawings 18, 61
Assembly sections 62
Associative 45
Auxiliary view 16

B

Base feature 43
Basic size 68
Bilateral tolerance 69
Bill of materials 19
Blind hole 60
Blocking-in 26
Bolts 61
Boring 71
Break lines 52, 58
Broaching 71
Broken-out section 58

C

CAD 6, 24
Centerlines 52, 55
Chain dimensioning 70
Child 45
Circle 25, 30, 54
Circularity 76
CNC machine tool 42
Computer-Aided Design and Drafting 6
Computer-Aided Engineering 6
Computer-aided drafting 6
Concentricity 77
Conceptual design 7
Concurrent engineering 7
Constraint-based modeling 43
Constraint-based solids modeling 41
Construction lines 26
Constructive solid geometry 40
Contour dimensioning 53

Crest 60
Cross-section 41, 43
Cutting plane 56
Cutting plane lines 52, 56
Cylindricity 76

D

Datum dimensioning 70
Datum plane 77
Datum plane symbol 77
Descriptive geometry 5
Design intent 43, 45
Design process 7, 17, 23
Design specification 7
Detail drawings 18
Detail view 59, 62
Diameter 54, 61
Dimension lines 52
Dimension placement 52
Dimensioning 55
Dimensions 52
Dimetric projection 13
Drill sizes 72
Drilling 71

E

Engineering graphics 2, 3
Exploded view 3
Extension lines 52, 55
External thread 60
Extruded 38
Extrudes 43
Extruding 41

F

Fasteners 59
Feature control box 76
Feature symbol 75
Feature-based modeling 43
Features 43
Finite Element Analysis 42
First-angle 49
First-angle projection 16

Flatness 76
Forging 73
Form tolerances 76
Freehand sketches 7, 17
Freehand sketching 13, 23, 24
Full section 56

G

Gages 67
Gaspard Monge 4
General tolerances 69
Geometric Dimensioning
    and Tolerancing 74, 75
Glass box 14
Graph paper 30
Grinding 71

H

Half section 57
Hidden lines 39, 51
History-based modeling 45
Honing 71

I

Ideation 7, 17, 23
Igor Sikorsky's 23
Inch dimensions 53
Integrated product
    and process design 7
Interchangeable parts 68
Internal thread 60
International Standards
    Organization 50
Isometric 12
Isometric freehand sketches 27, 30
Isometric grid paper 32
Isometric projection 26
Isometric view 5, 18
Ivan Sutherland 6

J

Jigs 68

L

"Landscape" orientation 50
Lapping 71
Lay 73
Leader lines 52, 55
Least material condition 80
Leonardo da Vinci 2, 3
Limit dimensions 69

M

Major diameter 60
Manufacturing process 71
Maximum material condition 79
Metric threads 61
Millimeter dimensions 53
Milling 71
Mind's eye 1, 23
Minor diameter 60
Model 37
Multiple features 55
Multiview projections 11, 14

N

Nominal 61
Nominal size 68
Nuts 61

O

Oblique freehand sketch 26, 28
Oblique projection 13, 26
Offset section 58
Orientation tolerances 77
Orthographic projections 4, 14, 32
Orthographic views 18

P

Paolo Uccello 3
Parallelism 77
Parametric 44
Parametric modeling 44
Parent 45
Parent-child relation 45
Parts list 19
Permanent mold casting 73
Perpendicularity 77
Perspective 13
Photorealistic 26
Pictorial perspective 3, 13
Pitch 60
Plus and minus 69
Polishing 73

Positional tolerance zone 79
Positional tolerances 78
Precedence of lines 52
Primitive modeling 40
Pro/ENGINEER 6, 42
Profile tolerances 76
Projections 11, 12, 14
Punching 71

R

Radius 54
Rapid prototyping 42
Reaming 71
Receding construction lines 27
Rendering 40
René Descartes 4
Revolving 41
Robust design 68
Root 60
Roughness 73
Roughness average 73
Roundness 76
Runout 77

S

Sand casting 73
Sawing 73
Scale 51
Schematic representation 60
Screw threads 60
Section views 56
Section-lining 56
Shading 40
Simplified representation 60
Single limit dimension 69
Sketch 41
SKETCHPAD 6
Solids modeling 6, 40, 42
SolidWorks 6, 42
Specification 7
Stamping 71
Standard size 50
Straightness 76
Surface finish 73
Surface model 39
Symmetry 77
Symmetry lines 52

T

Target size 68
Third-angle 49
Third-angle projection 15

Thread series 61
3-D model 39
3-D projections 11
Title block 50
Tolerance 53, 68, 70
Tolerance accumulation 70
Tolerance stackup 70
Tolerancing 68
Trimetric projection 13
True-position dimensions 78
Turning 71
2 1/2-D model 38
2-D model 37

U

Unilateral tolerance 69
Units 53

V

Visible lines 51
Visual thinking 1

W

Walter P. Chrysler 1
Waviness 73
William Farish 5
Wireframe model 39
Witness lines 52
Working drawings 17, 19